HILTON CARTER

WILD CREATIONS

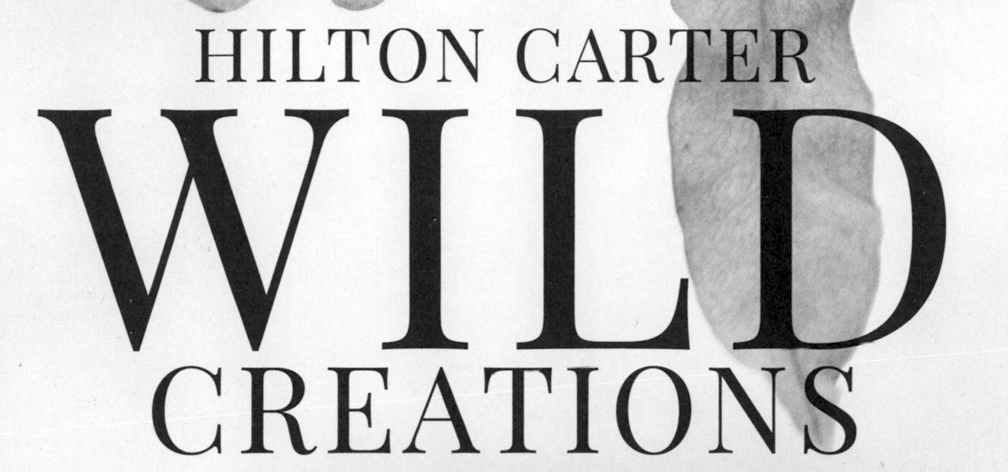

HILTON CARTER

WILD CREATIONS

INSPIRING PROJECTS TO
CREATE PLUS PLANT CARE TIPS
& STYLING IDEAS FOR YOUR
OWN WILD INTERIOR

CICO BOOKS
LONDON NEW YORK

Senior designer Megan Smith
Photographer Hilton Carter
Editor Martha Gavin
Art director Sally Powell
Head of production Patricia Harrington
Publishing manager Penny Craig
Publisher Cindy Richards
Cover and chapter opener art
Drury Bynum

Readers should make sure they are using the correct fixings and have tested the weight-bearing capacity of their walls and ceilings before hanging heavy plants and containers, or employ a professional to do so. The author and publisher cannot be held responsible for any damage occurring as a result of hanging the projects in this book.

Published in 2021 by
CICO Books
An imprint of Ryland Peters
& Small Ltd
20–21 Jockey's Fields
London WC1R 4BW
and
341 E 116th St
New York, NY 10029

www.rylandpeters.com

10 9 8 7 6 5 4 3

A CIP catalog record for this book is available from the Library of Congress and the British Library.

ISBN: 978-1-80065-025-1

Printed in China

CONTENTS

INTRODUCTION

People say things come in threes or that the third time's a charm. I believe that to be true. I mean, why not? I am from Baltimore, a place coined with the name "Charm City." So, that being the case, after making *Wild at Home* and following that up with *Wild Interiors*, I felt it was only right that I create a third book. The trifecta, if you will. In *Wild at Home* I spoke about the best ways to style greenery in a home, while also giving tips on how to care properly for your plant friends. In *Wild Interiors*, I took readers around parts of the globe, to be inspired by the journeys in greenery of individuals just like them, those in the plant-loving community, while also giving my thoughts on the best places in the home for particular types of plants to live and thrive. In *Wild Creations*, I want to give you, the reader, an opportunity not just to be inspired but to take action. It's in being inspirational that I want to go deeper. I don't want those reading this book to just feel excited or energized by an image, I want them to feel moved in a way that forces them to put down the book and go create the thing that they just saw or go tend to their plants with the care they'd give to any living thing in their home. That's real inspiration. The power to awaken an individual's creative sleeping giant and have them feast on the opportunities in front of them. That's the direction the next 200 pages should take you toward.

Wild Creations is a book I needed to create because it's a book that I selfishly wanted for myself. As a plant and interior stylist, there have been times when I have an idea for how I'd like to utilize a plant in my home, but there hasn't been a product that reflected my sensibility or style, and it's made me want to just create it myself. For example, I knew I wanted to have a terrarium in my home but wanted it to be different than any I had seen before, so I purchased a fillable glass lamp and turned it into a terrarium. I could feel how special that was and the statement it made in a room was evident. Honestly, it's in those moments when I know I particularly want something and I believe others might want it as well, that I know I'm on the right path. While I'm a lover of plants, I'm just as much a lover of interiors and style, and my aim has always been to live in that space where the two come together. Therein lies the beauty of bringing the outdoors in.

WILD
INTERIORS

The first chapter, WILD IDEAS, is the heart of this book. It will give readers a chance to show a little of their creative side, getting their hands in the mix, to create beautiful projects in the comfort of their home. I think right now, many of us are looking for ways to create with our hands and literally do it ourselves. With ideas from how to create a concrete planter to a detailed breakdown of how to paint a jungle mural one tone at a time, each project is designed to help make your space feel more lush, stylish, and, of course, alive. My plant hammock project, which I shared in *Wild at Home*, is a good example of this and because there was such a great response to that part of the book, I wanted to do more of it. The projects shared throughout this book are all ideas that I've been excited about for a while now or concepts that I've been thinking about, and now finally have the opportunity to put out into the world. In the same way that the at-home chef uses their favorite cookbook to spice up their plate, I want *Wild Creations* to be that to the plant and interior design lover.

In the chapter WILD HACKS, it felt important to share the pro-tips of sorts that I've come to learn over the years. These are simple ideas to help you care for and style the plants that you are bringing into your interior space, which maybe you haven't thought of. Did you know that you can use No.2 (HB) pencil shavings to help repel bugs? No? Well, you can. That's what I mean by simple. We will discuss ideas from how best to pack your plants for a move, to how to style the surface of a plant's soil, and just cool/inventive ways to set the vibe. The goal for these hacks is to bring a little fun to your weekly care routine or alleviate some of the stress when it comes to the plant issues you might encounter.

In the chapter WILD RANTS, I do just that... rant. Wild Rants could be its own book when I think about it. I refer to this section as rants because I feel like it's me on top of my soap box screaming about my love for plants and how transformative they can be in your life, and I just wanted to put that energy out into the world. After two books, I felt driven to drill home a few thoughts that I've had over the years from discussions with those in the community or messages sent my way. I fall into my feelings and go from discussing the true power of light to the importance of getting a plant sitter. And let me tell you, it's important. I feel passionately about the relationship we have with green life and wanted to help others out there have a better understanding of why it's important to make that connection. While many have been running out to fill their homes with as many plants as they possibly can, I hope this chapter will be a PSA (Plant Service Announcement) on why we should be thinking more deeply about plant life and the goodness it can provide. I would certainly be remiss if I didn't talk a little about some of my favorite indoor plants and how to care for them. I mean, who would do that? I've covered some of my top plant friends in *Wild at Home* and *Wild Interiors*, so it's only fitting that I do that in the WILD PLANTS chapter of this book, too.

I find myself so overwhelmed with excitement that I've gotten the opportunity to share my passion with the world once more. This plant-loving community has grown so much over the past few years and I feel fortunate to be amongst all of you. Whether you're a novice or someone that considers yourself to have two green thumbs, eight green fingers, and ten green toes, I think *Wild Creations* has something special for you. So grab your apron and gloves, and let's get WILD!

WILD IDEAS

This book here, the one that you're reading, was the initial Wild Idea. I thought, how do I get my publishers to let me make a book filled with DIY projects that my friends in the plant community and I would be excited about? So, I jotted down a few concepts, sent them over to my publishers, and you know what... they totally decided to pass on it. They thought it would never work. OK, clearly I'm kidding. You're reading the book, so obviously they were into the idea. Really into it. With the COVID-19 pandemic affecting so many of us across the world, I wanted to have some ideas in this book that would encourage those stuck at home to get inspired and creative. If you're anything like me, finding ways to refresh my living space brings me joy. So when I can find ways to refresh my surroundings that involve introducing greenery, it's euphoric. Biophilic design is moving the world into the future, so ideas that help merge indoor and outdoor are extremely vital. In this chapter of the book, you will find DIY projects that are close to my heart, some that will turn your home into a jungle without using living things, and some that will truly bring out your inner child. But I guarantee, all will get your hands dirty!

While we might not all have the same artistic talents or access to materials, I tried my very best to make sure these projects could be created by everyone, from all across the world. Many of these projects can be done alone, but some will require you to ask for a little help. For me, there's nothing more fulfilling than working on these projects with family or, if and when possible, friends. Each project breaks down what materials you'll need and, more importantly, how long each project should take you, making it easier for you to manage. Enjoy!

TEQUILA
BLANCO
MEXICO

LIVING ART: HOW TO MAKE A WALL MOUNTED PLANT

So many people today are looking for creative ways of blurring the line between indoors and outdoors. For me it's always about what plants would thrive in the light provided in the areas I would like to place them while still showcasing my own individuality and style. As an interior and plant stylist, there is real artistry in creating the ideal gallery wall. Of course, this all starts with selecting the right art. And in doing so, it's important to make sure you have a good mix of sizes and shapes when it comes to framing, and, of course, dimensions to make things stand out.

The best gallery walls I've seen have all had one thing in common: variety. They speak about who the individuals are that live in the space, yet also create a bit of a wow factor. I've always been in search of the "wow" and finding ways to sneak in a bit of greenery and tie in plant life to your gallery wall can take you there. Listen, with the amount of plants I have in my home, believe me, I know how quickly you can run out of floor space. But if you have the light and the time to care for more plant friends, why limit yourself to horizontal space only? When possible, go vertical! In my first book, I used this idea with the plant hammock above my bed. Not only are you finding more room to display your love for plants but you're also creating living art. So I wanted to share with you how to make a smaller piece of living art by creating a wall-mounted plant.

If you're a plant lover like me, you might have seen a mounted staghorn fern in your visits to plant stores. But have you ever considered how to create and style your own mounted plant? Well if so, today I'll show you how to create that living art and how to make your gallery wall feel a bit more alive.

PROJECT TIME
1 HOUR

WHAT YOU'LL NEED

A mountable plant: a staghorn fern (*Platycerium bifurcatum*), *Philodendron*, or *Hoya* are good choices.

Preserved sheet moss (*Hypnum cupressiforme*) and reindeer moss (*Cladonia rangiferina*)—the amount you'll need will depend on the size of your plant and the wooden board you use, but for the one in the photos I used an 11 x 8in (28 x 20cm) piece of sheet moss and a few handfuls of reindeer moss.

Wooden board—I used some reclaimed wood slabs, but a new cut of wood from a home improvement store, a plaque, or a piece of driftwood would work just as well. Just understand that the larger the piece of wood you start with, the more room you will give your plant to grow without having to remount it.

Twine or fishing line, approx. 36in (90cm)

Large sawtooth picture hanger, to support at least 20lb (9.1kg)

2 x ½in (1.2cm) nails for picture hanger

Approx. 8–14 x 1in (2.5cm) nails (depending on size of board and plant)

Sharp scissors

Tape

Pencil

Hammer

Bowl of water

Possibly a friend to help!

1

Nail the picture hanger into the back center of the board. Cover the hanger with tape so that it doesn't scratch your work surface. Trust me, I've learned the hard way. Then flip the board over and place the potted plant on the front, and decide where you want to position it. Use a pencil to draw a circle where you'll be mounting your plant.

2

Take some of the reindeer moss and place it over the circle you've drawn on your board until it's covered. Now that you have the correct amount that you'll need, remove it from the board and place it in a small bowl full of water. Set it to the side.

3

Hammer 8–14 of the 1in (2.5cm) nails (depending on the size of your plant) into the board around your circle, about a ½in (1.2cm) wider than the widest part of the plant pot. Angle the nails slightly outward and upward toward the edge of the board.

4

Place a layer of damp moss between the nails. Build it so it's thinner at the top and thicker at the bottom. Remove the plant from its pot, loosen the roots, and remove some of the excess soil.

5

Place the plant firmly on the board and, picturing how much you'll need to cover the soil of your plant, cut pieces of the sheet moss, wet them, and then wrap them around the soil, pressing and shaping the moss and soil between the nails.

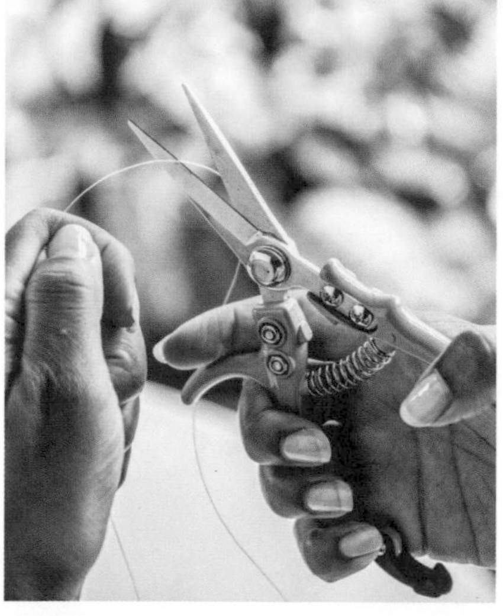

6

Have a second pair of hands to hold the plant in place while you secure it. Cut a 36in (90cm) piece of fishing line or twine and tie it to a nail, leaving a 3in (7.5cm) tail—you'll use this to tie off the string when you're done securing the plant.

7

Pull the line tightly across the plant, hook it around a nail, and cross over to another nail, pulling the line taut to secure the plant. Keep going until you've hooked each nail once or twice and the plant feels secure.

I used fishing line for this staghorn so you can see what the plant should look like when it's secured. Trim the excess line and hang it up!

STYLING INSPIRATION

When it comes to styling your living art, the idea is to surround this 3D work of art with 2D pieces so that it stands out, not only in terms of its presence but also spatially. In most cases, your mounted plant will be the statement piece on your gallery wall, begging for all the attention. And believe me, it'll get it. So, finding the perfect balance of art to go around it is important. Have fun!

8

Water your plant immediately after mounting. Follow the care instructions for your specific plant, and keep in mind that mounted plants dry out faster than potted plants. As a general rule of "green" thumb (see what I did there?), mounted plants do well in bright, indirect light with watering every one to two weeks. When it comes to watering a mounted plant, there are two ways to keep your living art thriving. One, mist at the root ball (moss pouch) once a week. Two, soak the entire plant, including the wooden board, in a sink or tub of lukewarm water for about 10–15 minutes once a week, making sure the root ball and foliage are submerged. Don't forget to let it dry before placing it back on your wall. For more care tips on staghorn ferns, see p.228.

KSM CANDLE CO.
No.35 Gathered
WHITE FANG · JACK LONDON

THE "GATHERED" CANDLE WITH KSM CANDLE CO.

PROJECT TIME
45 MINUTES

What is it about bringing candles into a space that can truly make it feel like a home? Maybe it's being able to watch as that burning wick warms up the space or the aroma tints the air. While NASA has done studies on how plants can add oxygen to a home and clean the air, and even with over 200 plants in our home, the sense of clean air definitely comes from the candles and incense we burn throughout. They create a calming aroma that moves like a creeping fog from room to room. I don't know about you, but I find myself being drawn to the flame. Maybe it's because I'm a fire sign or the fact that there is something so mesmerizing about watching a flame dance around in the space, swayed by the indoor air. Burning candles in the home doesn't just help to add a fresh smell to a room, they can also help to trigger a memory, or ease the stress of the day.

When thinking about creating the perfect vibe in my home, I thought why not make my own candle with a scent that reflected everything about my idea of "home." So I reached out to my friend Letta Moore, who owns KSM Candle Co., a candle-making company here in Baltimore, to help me craft the perfect scent. Letta's vision wasn't to create just another candle company, it was "to create a company where people could get together and share their passion for creating." That mission statement shines through if you're one of the many that have attended her candle-making workshops. That idea of making a place where people could come together lit a spark in me, that lit the wick to burn the candle of my own idea of home. With home being the place

that we all gather together with those we love, I needed the scent of the candle to create a memory of home for me. And with my home being about bringing the outdoors in, I wanted a floral candle that could enhance the blooms of the flowering plants inside throughout spring and summer. Letta pulled together some floral scents for me to pick from and the perfect combination was a mix of magnolia, hibiscus, and honeysuckle. Letta put her professional nose to it and agreed. Once we had the scent, I knew exactly what we needed to call the candle... "Gathered." Not only does it represent gathering in the home with your loved ones, but also the gathering of thoughts when one is alone, or gathering the contents to make an idea come to life.

So we wanted to share with you how to make your own "Gathered" candle at home. Most of the things you'll need to make this candle can easily be found online, but you can also find it all at ksmcandleco.com and you can find Letta on Instagram at @ksmcandleco.

WHAT YOU'LL NEED

Approx. 16oz (450g) soy wax flakes

½fl oz (15ml) mixture of magnolia, hibiscus, and honeysuckle fragrance oil, or your own special chosen scents

Vessel (old Mason jars are great)—please make sure the vessel can be used for a candle. No wood and no plastic. Ceramic, cement, and tin should all be sealed with a fire-resistant sealer before use.

Wick, at least double the height of the vessel (if you are using soy wax, soy-coated wicks are best)

Double-sided wick sticker or glue to fasten wick to the bottom of the vessel

Something to hold the wick in place—this could be a number of things, such as a clothespin (clothes peg), stick, or a chopstick

Microwave-safe container to melt the wax in—if you don't have a microwave to melt the wax, you can use the double boiler method: place the wax flakes in a glass bowl or metal pan that you won't use again for food and heat over a pan of simmering water until melted.

1

Pour the wax flakes into a microwave-safe bowl. Choose something you don't plan to use for food again. An old Tupperware container with a missing lid works great.

2

Microwave the wax for 30–60 second intervals until completely liquid, stirring between each melting session. Don't forget that the contents will be hot, so use proper caution. That's just a nice way of saying "Don't use your hands."

While the wax is melting, you can prepare your jar for pouring (see steps 3 and 4).

3

Remove one side of the double-sided wick sticker and place on the bottom of the wick (the metal side), or use glue here.

4

Remove the other side of the sticker and center it inside and on the bottom of your jar, making sure to press down firmly so it's nice and stuck.

Wrap the wick tightly around the clothespin, stick, or chopstick, while gently pressing the coil in place for 10 seconds. Your natural body heat will help hold the coil in place so that it doesn't unravel when you let it go.

5

Once the wax is fully melted, it will be translucent. If not, place it back in the microwave for an additional 20 seconds. Take the wax out of the microwave, add the scent oils, and stir for 2 minutes.

6

Pour the wax slowly and carefully into your container.

Allow your candle to solidify at room temperature on an even surface for about 20 minutes or until solid.

7

Once your candle is completely solid, take a sharp pair of scissors and trim the wick to ¼in (5mm).

CANDLE BURNING TIPS

Before I worked with Letta on this project, I wasn't aware that candles form a memory. A memory? Exactly. What?! Basically, every time you use them they remember how they were previously burned, so it's important to create the right memories. I guess we can learn a lot from candles, huh?

Light your candle for as long as it takes for the entire top layer of the candle to melt completely. You want the melt pool to be about a ¼in (5mm) deep around the jar. This will do two things for you: it'll make your fragrance as strong as it can be for that candle and every time your candle solidifies, it will solidify evenly.

Trim the wick. This will help to reduce soot. The recommended height is ¼in (5mm).

KSM CANDLE CO.
No.35 Gathered
SOY CANDLE
OZ
HOURS
RESPONSIBLY SOURCED IN THE U.S.

LEATHER HANGING PLANT STAND

When speaking about plant styling, I frequently talk of creating levels in a space using greenery to enhance the effect. Having plants above, at eye level, and below is reminiscent of what you would encounter in nature, so creating these levels in your home is the most seamless way to blur that line between indoor and outdoor.

In the case of having plants above eye level, the easiest way to achieve this is with hanging planters. Hanging plants throughout a home can make a room feel more lush and move one's eyes around the room to fully take in the environment. For the viewer it's a joy, but for those caring for these hanging plants it can become a bit of an inconvenience. This is mainly due to the fact that most hanging planters have drainage holes, so you can't just water them where they hang. To properly care for them you have to let them drain out, and without anything to catch that runoff water, you'd get a pool of water building up on your floor below. No one wants that. To do the job properly, you'll have to get up a ladder to bring the plants down to water them, let them drain in your sink or tub, and then, once dry, return them to their spots. The more hanging plants you have, the more work it takes to water them. During the days that I water my plants, this process is what eats up most of my time. I thought, if only these hanging planters could have a wide base tray under them so that you could just go up a ladder and water your plant where it hangs, allowing for enough runoff water to sit in the tray below the pot but also not taking away from the beauty of the planter and the plant.

PROJECT TIME
1 HOUR

WHAT YOU'LL NEED

Materials

3 x 4–5oz (1.6–2mm thick) vegetable-tanned leather straps, 48in (120cm) long x ¾in (2cm) wide (you can find leathers online or at any hobby or craft store)

2 x O-rings, size 1½in (38mm)

4 x D-rings, size 1in (25mm)

14 x small double-cap rivets with a cap and post size of ¼in (6mm) Note: If using leather heavier and thicker than 4–5oz (1.6–2mm), size up the rivets to a post size of ⅓in (8mm) or larger with a post size of ½in (13mm) depending on thickness.

Tools

Scissors

Tape measure

Ruler

Dull pencil

Hole punch

Rivet setter

Rivet anvil

Hammer

Sponge

Bowl of water

Cutting board

Note *The measurements and materials listed above are for the example shown in the photos, but this project is very customizable. You can use vegan leathers, upcycle thrifted belts, or even dye or paint the leather. The one thing to remember is just be creative and make it your own.*

I had seen many hanger planters in the past that could hold a base tray under the pot, but they would fully wrap around the pot, tray, and plant, not allowing for them to breathe and show off their beauty. I wanted something that propped the plant and planter up high in my home, but still allowed them to show off. What I wanted was basically a hanging plant stand. Something not only functional, but stylish.

I sketched out some ideas of how this hanging plant stand could look and sent them to my friend Sara Tomko, who works in leather design, asking if she could help me create something beautiful to display my hanging plants. I made it clear that I wanted to be able to place various sized base trays inside of it. Sara loved the idea and a few days later she had the prototype created. It turned out so great that I'm sharing it here, so you can make one yourself.

LEATHER BASICS

Grain side *The decorative side of the leather that will be the visible outside of your hanger*

Suede side *This is the opposite side of the leather that is textured and will be the inside of your hanger*

Vegetable tanning *An eco-friendly tanning process that uses natural materials to turn cowhide into leather*

Casing leather *The process of adding water to vegetable-tanned leather to make it softer and easier to cut, hole punch, and shape*

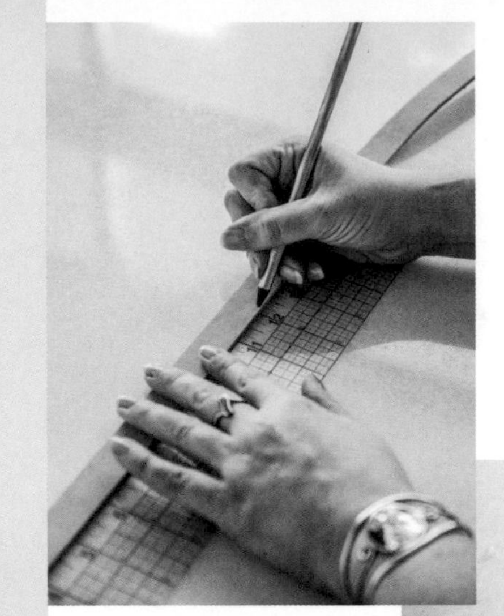

1

Cut one of the 48in (120cm) leather straps into four 12in (30cm) lengths. These will be the straps that hold the base tray.

2

With the sponge, soak up some water and wipe down each 12in (30cm) strap. Wetting the leather makes it soft and durable. Be sure to wet the entire strap to avoid uneven discoloration. Sponge the leather until it cannot hold any more water. It's now ready to mark and hole punch.

3

With the dull pencil and ruler, mark one end of each 12in (30cm) strap with two points at ¼in (5mm) and 2in (5cm) in the center of the strap. This is the end that will connect to the O-ring at the bottom.

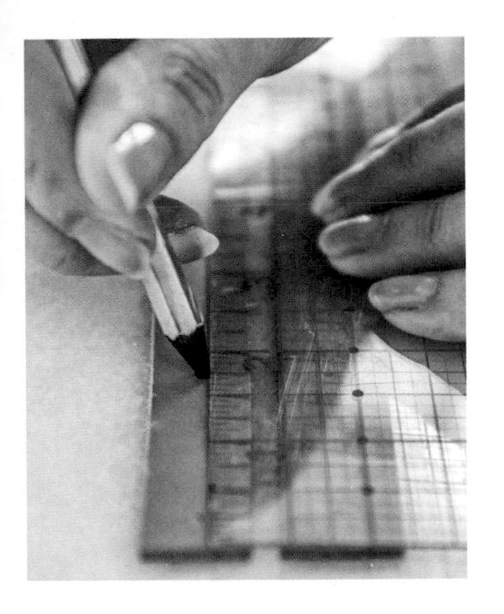

4

At the opposite ends of the 12in (30cm) straps, mark two points in the center of the straps, ¼in (5mm) and 1¾in (4.5cm) from the end. This is the end that will connect to the D-rings.

Now sponge down the remaining two 48in (120cm) leather straps.

5

With the tape measure, make a mark at the 24in (60cm) halfway mark. With your ruler make a mark 1in (2.5cm) from each side of the 24in (60cm) halfway mark in the center of the strap. Now mark the ends of the 48in (120cm) straps with two points at ¼in (5mm) and 1¾in (4.5cm) from the ends, in the center of the strap. These ends will also connect to the D-rings.

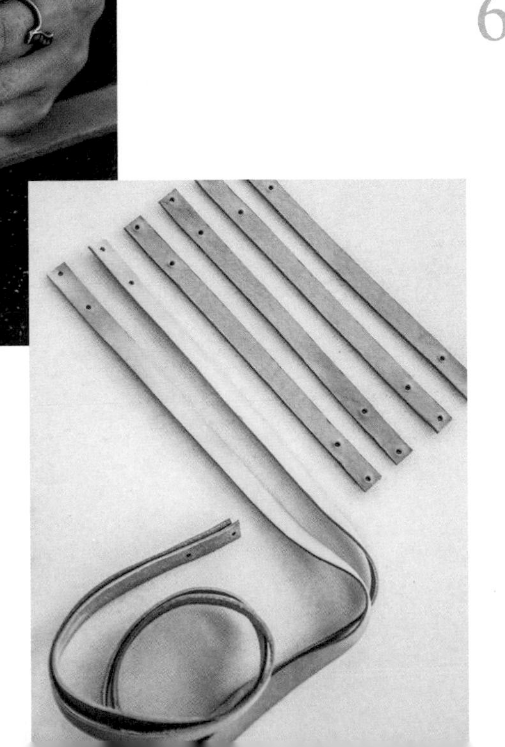

6

Now it's time to punch the holes into all the leather straps. Grab the cutting board, hammer, and hole punch. Place one 12in (30cm) leather strap on the cutting board and punch all four holes you marked with the hole punch and hammer. The wet leather will make punching holes much easier. If your leather has dried a little you can re-wet it with the sponge. Repeat this step with the three remaining 12in (30cm) straps. Now punch all six marked holes in the two 48in (120cm) straps.

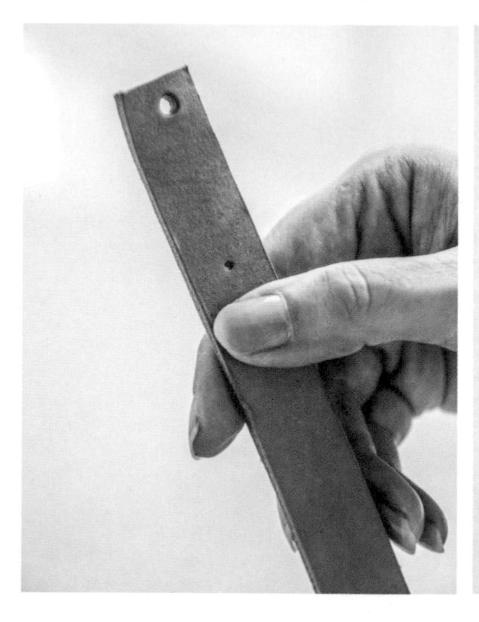

7

This is the time to bring it all together, but first let's organize the straps. Grab the 12in (30cm) straps and one O-ring. Line up the straps so the ends marked at ¼in (5mm) and 2in (5cm) from the end are at the bottom. This is the end that will connect to the O-ring.

8

Place the O-ring on the cutting board. Take one 12in (30cm) leather strap, suede side facing up, and place the O-ring marked end under the O-ring. Wrap the strap around the O-ring so the two newly formed holes meet. The long end of the strap should face away from the O-ring.

9

You will need the rivets, rivet setter, and hammer. Double-cap rivets have two parts: post side and cap side. Insert the post side from the bottom through the two holes. Place the cap end of the rivet on the other side. Squeeze the rivets together to snap them shut. Now, using the rivet setter, rivet anvil, and hammer, position the rivet on the anvil so the bottom cap lays inside the dip on the circular portion of the anvil. Place the concave portion of the rivet setter on top and give it a good tap with the hammer. Repeat steps 8 and 9 with the three remaining straps around the O-ring. Be sure to check that you are connecting the right end to the O-ring.

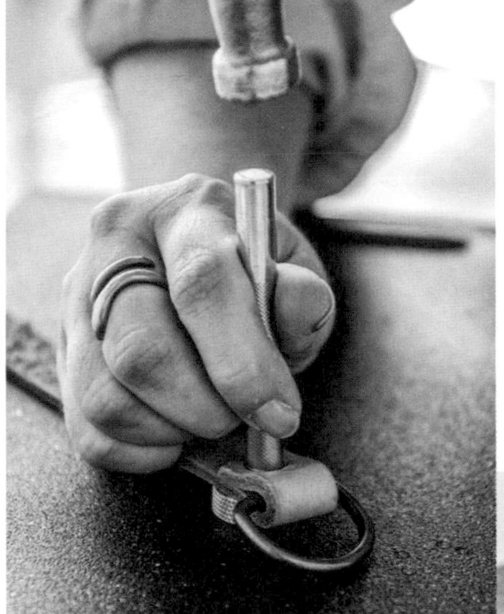

10

Using the D-rings, repeat the riveting process for the opposite ends of the straps. Now the straps that will hold your base tray are complete.

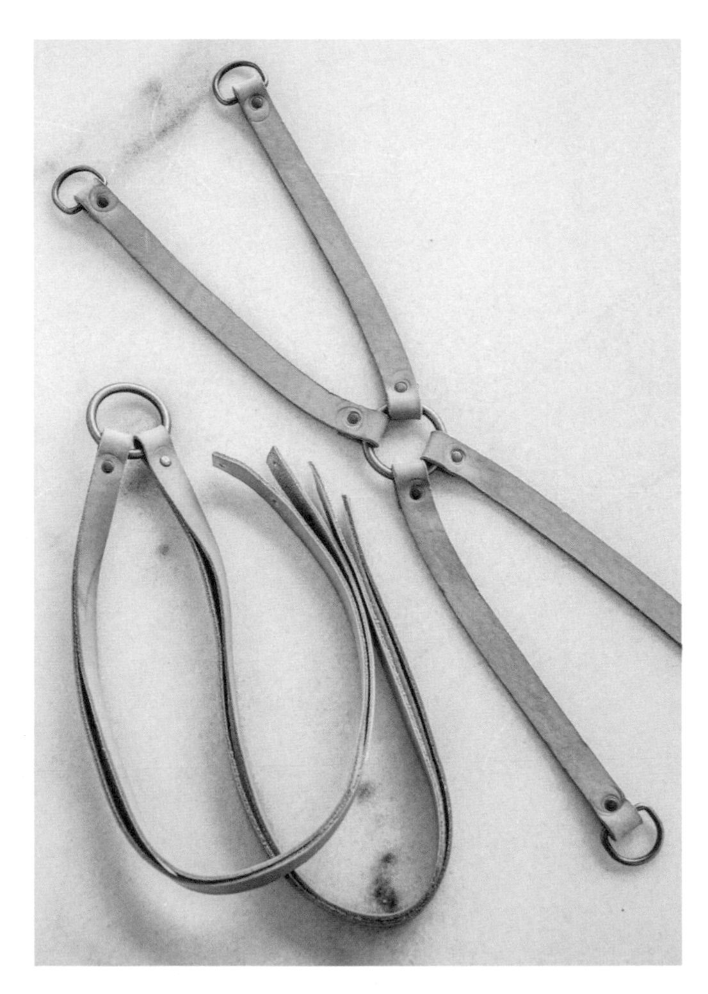

11

The last part is to connect the long straps. Feed one 48in (120cm) strap through the remaining O-ring. Rivet the two holes you made at the halfway point to secure the straps to the O-ring. Repeat this step with the other 48in (120cm) strap.

12

Almost done! Now we need to connect the long straps to the short straps. Rivet the ends of the 48in (120cm) straps to the D-rings of the 12in (30cm) straps. This part is a little tricky. Make sure to match the leather grains as you attach, as it can be easy to attach them backward—you want the suede side facing inward. The straps should connect at the four points in a circular fashion so a base tray will sit inside and be supported. Avoid crisscrossing the long straps to the short straps while attaching.

And you've done it! You now have your own plant hanger to display your beautiful plants and also your beautiful planters!

MIX & MATCH POTS & BASES

When matching pots and base trays, I like to consider the color, texture, and size of the pot first and then place the best base tray underneath. Lately, pots have started to come with base trays, but this doesn't mean you have to stick with that tray. Feel free to get a little creative and original by purchasing two pots and swapping their trays, or using old plates or trays to make them stand out.

You can find Sara Tomko, designer and creator of Hide and Peak leather goods, at hideandpeak.com and on Instagram at @hideandpeak

CONCRETE PLANTER

The moment the first person thought to bring plants indoors to make their home feel less stale and more interesting, they must have figured out that they needed the soil and roots to be contained in something. I mean, they wouldn't just build mounds of soil in the corners of their living room or bedroom and call it a day, would they? Or even if they did, at some point, after continuously sweeping up soil, they would have decided that there had to be a better way. And, of course, that was to place the plant in a planter. Since that day, the plant and the planter have gone together like bread and butter, or cereal and milk. You just can't have one without the other. I like to call the planter the "dress" or "trousers" of the plant. While the plant itself should command most of the attention, it's the planter that gives it a little individuality. The planter is where the stylist gets to have fun and allow its color, shape, and material to connect with the space that surrounds it.

With houseplants becoming so popular over the past few years, what's also had a rebirth during this time is design in planters. There are so many amazing planters out there to dress your plants in that it can become a bit overwhelming trying to pick out the perfect additions for your home or space. While I love being able to mix and match a little of the old with a little of the new, I also like getting my hands a bit dirty and creating my own. So I thought, why not share how you can all create your own planter at home. With my love for brutalist architecture, I wanted to

PROJECT TIME
28 HOURS (2 FOR CONSTRUCTION, 24 FOR DRYING, AND 2 FOR FINALIZING)

create a planter made of concrete. Concrete is a material that has been used for such a long time when it comes to design because of how inexpensive, durable, and easy to manipulate it is. While it may not be the most lightweight material, it's something that will last you for a very long time if taken care of properly. For me, I just love its texture and color, and when placed next to plants as a planter or even just in design, the way the two work to vibe off each other is ethereal. The balance of cold and warm, and light and heavy, does so much. I want to share with you tips on how to create your own concrete planter.

WHAT YOU'LL NEED

This will make a 7 x 7in (18 x 18cm) planter.

Ruler

Spoon to mix

File for sanding

Sandpaper in 2 different grits (120 and 220)

Measuring cup

Box cutter with sharp blades

Pencil

Adhesive tape

Medium plastic tub

3/16 x 24 x 36in (5mm x 61 x 91.5cm) foam board

Cutting board

Rubber gloves

Cooking spray

Mask

Safety goggles

Drill with ¼in (6.5mm) drill bit

Sponge roller

40lb (18kg) of high strength concrete mix—it took 14 cups (approx. 8kg) of concrete to make my planter.

Concrete sealant

Spirit level

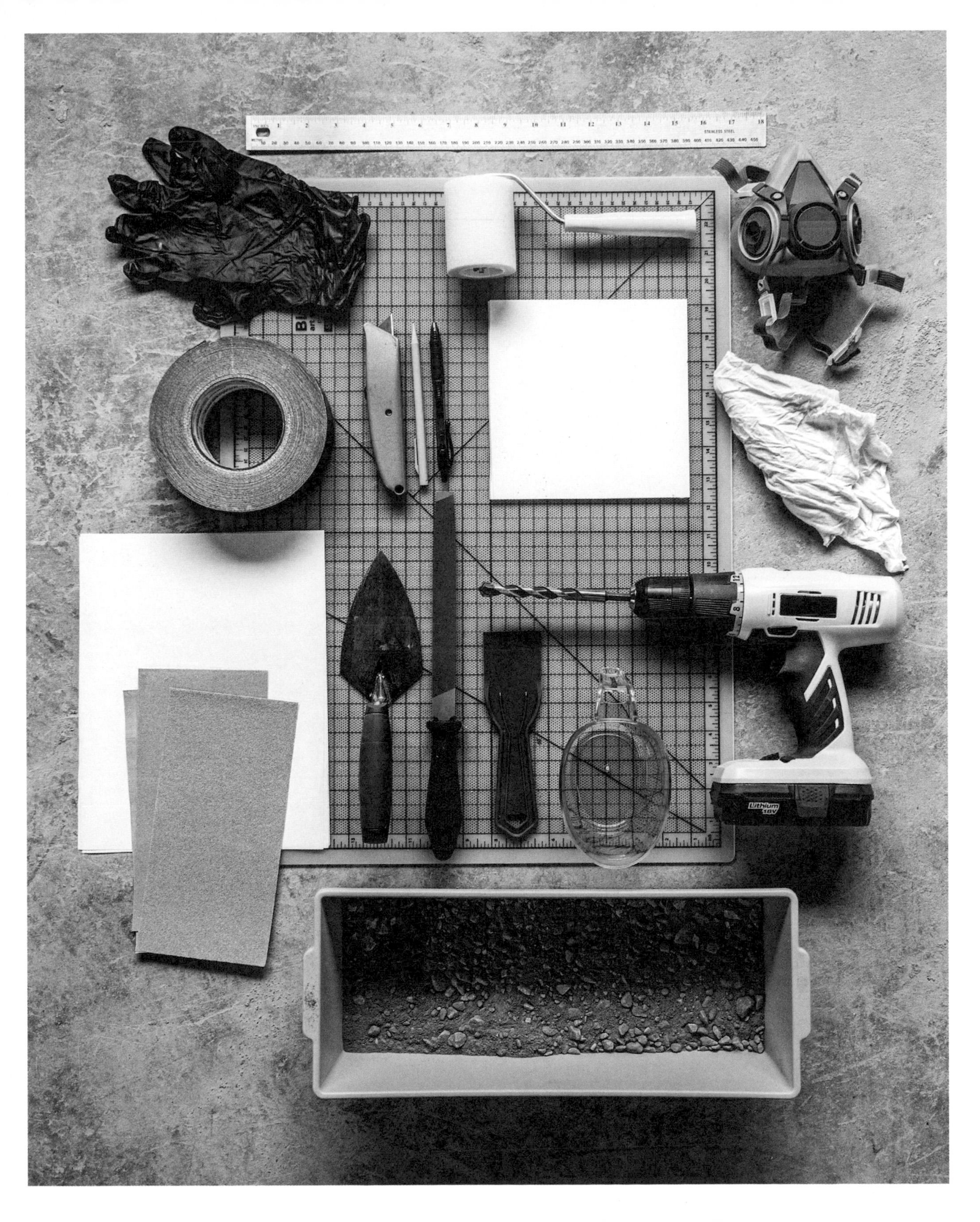
Lithium
18V

1

Draw out the size of the planter you want to make. I wanted a 7 x 7in (18 x 18cm) planter, so the first thing to do is draw it out to make sure you're creating the right size pieces. To make your planter the size you want, you'll have to consider the thickness of the foam board. The foam board is 3/16in (5mm) thick and that number will play a large part in the masurements when cutting out your forms.

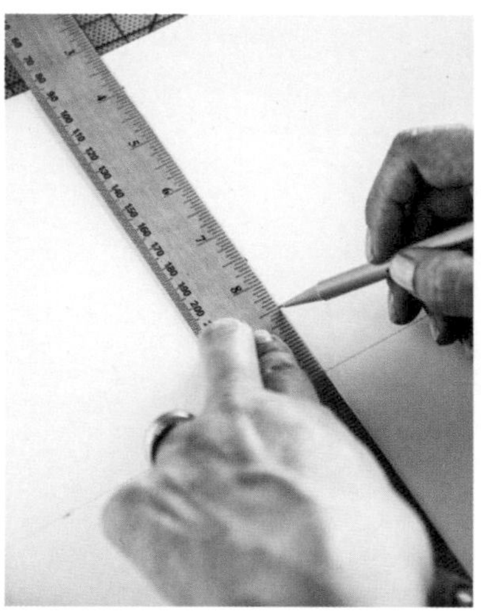

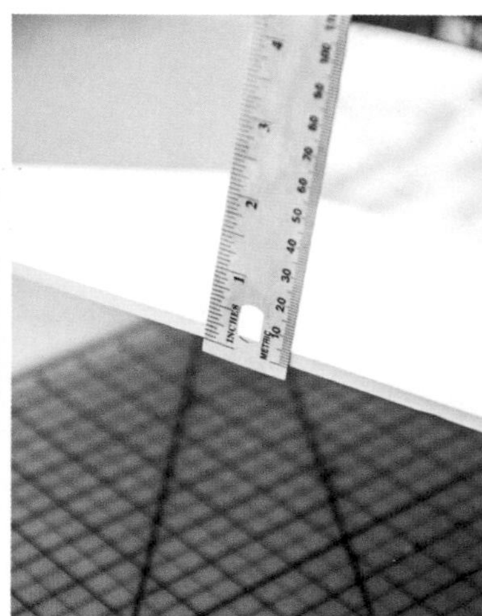

2

You need to create two forms. An INNER FORM and an OUTER FORM. To make a 7 x 7in (18 x 18cm) planter, you'll have to create a BASE for your OUTER FORM that is 7 6/16 x 7 6/16in (19 x 19cm). Remember you'll be pouring the concrete into this form, so the thickness of the foam (3/16in/5mm) in each side, once subtracted, will give you 7 x 7in (18 x 18cm). Next make the SIDES. With a new sharp blade, cut out two sides that measure 7in high x 7 6/16in wide (18 x 19cm) and mark them as 1A and 1B. Then cut two sides that measure 7 x 7in (18 x 18cm) and mark them as 2A and 2B. It's important that your blade is sharp as this will allow you to make clean and precise cuts.

Note When cutting the foam board, it's important to take your time. Don't try to make the cut all in one slice. Glide your blade along your line a few times and watch as the blade makes an even cut.

3

Take a marker and on the boards, mark what the boards are. For example, "INNER BASE." Now that the OUTER FORM boards are cut, it's time to make the INNER FORM. My goal was to make my planter ½in (1.2cm) thick. You can make yours as thick as you want, but I wouldn't make it thinner than ½in (1.2cm), so you don't have a planter that can easily crack. To make my planter ½in (1.2cm) thick, I made the INNER FORM 6 x 6in (15.2 x 15.2cm). Start by cutting out the base at 7in high x 6in wide (18 x 15.2cm). Next make the SIDES. Cut out two sides that measure $6\frac{13}{16}$in high x 6in wide (15.7 x 15.2cm) and mark them as 1A and 1B. Then cut two sides that measure $6\frac{13}{16}$in high x $5\frac{10}{16}$in wide (15.7 x 14.3cm) and mark them as 2A and 2B.

4

Tape the OUTER FORM together. Remember that the SIDES will stand on the top of the BASE, not to the side of the BASE. This OUTER box can be taped heavily but try not to have any tape make its way inside the foam form. Use the blade to cut off any extra tape. Repeat the same steps for the INNER FORM. Once both forms are taped, set them to one side.

5

Put on your rubber gloves. Take the plastic tub and fill it with 14 cups (approx. 8kg) of concrete. With the measuring cup, measure out 4 cups (1 liter) of clean water and slowly pour a little over the concrete. Stir together with the spoon. Add a little more water as you go. The goal is for the mixture to have the consistency of oatmeal. If it seems like it's still a bit too thick and clumpy, add a little more water.

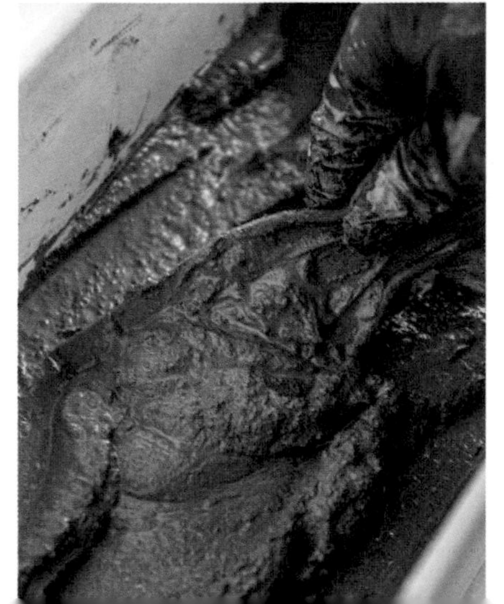

6

Take the cooking spray and spray the inside of the OUTER FORM and the outside of the INNER FORM. This will make it easier to separate the foam boards from the concrete.

7

Pour the concrete mix into the OUTER FORM. Fill it up at least halfway. Shake the form, banging it on the table to even out the concrete and remove any air bubbles. Now take the INNER FORM, center in over the OUTER FORM and press it down into the concrete mix. Press down until you have ½in (1.2cm) of the INNER FORM sticking out at the top. This is to make sure that the base of your planter is ½in (1.2cm) thick. You may need to use another piece of board and a brick or weight to hold the INNER FORM down in the concrete. Let dry for 24 hours.

8

Once the concrete is completely dry, use the blade to carefully cut away the tape. Remove the boards carefully to avoid damaging the planter.

9

Once the planter is out of the boards, you'll notice that some sides are uneven. Use the file to gently and slowly file the sides you need to even out. I used a spirit level to check my work. Then use the 120-grit sandpaper to smooth out the sides and finish with the 220-grit sandpaper.

10

Flip your planter over and drill a ¼in (6.5mm) hole in the center of the base. This will allow water to drain from the soil.

11

Use the lint-free cloth and wipe away any remaining dust. Then take the sealant and paint the sides, bottom, and inside of your planter. Please follow the manufacturer's instructions on the sealant for application. Let dry for at least 24 hours before placing a plant inside.

12

Add soil to the planter, followed by your choice of plant, and enjoy!

Now that you've gone through the process and made your own concrete planter, you'll probably want to make a concrete base tray to go with it. Follow these same steps, but make it shorter and wider. It's that simple. Once you get your hands mixed into concrete, I'm sure you'll want to make so many other projects out of it.

AT HOME
HILTON CARTER

CLAY BASE TRAY

The unsung hero in the indoor gardening world might easily be the base tray—well, at least in my opinion. The base tray lifts your planter and plant on top of its shoulders and catches all the excess water that comes seeping out the drainage hole of your pot. It might not get a lot of time in the limelight, but its role is important for the health of your plant and also that of your floors. I mean, you're going to want something under that planter to catch all of the excess water so it's not running down your hardwood floors or pooling into your Persian rug. Having a base tray will help you avoid that. The thing is, in many situations, your options for base trays are limited. The most common base trays out there are the standard terracotta trays made to go with their terracotta planters, or the plastic base trays that lack any style but get the job done when you have a planter that didn't come with its own base. And that last part there is the biggest issue right now.

Most planters don't come with a base tray, so you are forced to purchase plastic trays just to make it work. I've suggested hacks like using old dinner plates, quiche trays, and even large bowls to use as a base tray, because at least then you could bring in a little style, color, and uniqueness.

It's the ability to be unique that separates us from each other. While you may have purchased the same planter as someone else, the base tray you place underneath and how you style it can be completely different. And it's in creating that base tray that I think you'll

PROJECT TIME
60 MINUTES TO MOLD, 3 DAYS TO FULLY DRY

WHAT YOU'LL NEED

1lb (450g) air-dry modeling clay

Cutting mat or rubber/plastic board

Clay sculpting tools: spatula, single-wire end tool, detailing tool

Gloves (or feel free to use your bare hands)

Spray bottle

Pencil (just in case you want to sketch out any designs)

Sandpaper: 120- and 220-grit

Bowl or planter to use as a guide for the tray size/shape

Note These are the materials needed to create a 7½in (19cm) base tray, as shown. To make a smaller or larger base tray, adjust the amount of clay.

find fun and exciting ways to make your plant stand out. In this book I'm giving you the recipe to create your own planter, so I felt it was only right that I give you the recipe to create your own unique base tray. And to do that, what better way to bring the outdoors in than to use a material straight from the earth, and that's clay. Working in clay makes you feel connected in a small way to the earth, but mainly to your work. Your touch is in the work. In some cases, if you decide not to use gloves, your actual touch, your fingerprints, are forever captured in the clay. That's as unique as it gets. So, here are my steps to create your very own clay base tray.

STYLING INSPIRATION

When styling a base tray, think about the surface it's going to sit on. Terracotta and clay are porous, and will release a little moisture below the tray, so don't place them on wooden floors or windowsills that aren't sealed. I like to raise them off a surface using old books, old cutting boards, or pedestals. *Learn more tips in* Level Up, *on p.164.*

1

Take the clay and place it on top of a clean cutting mat or board. It's important that the surface is clean so you don't end up with unwanted debris in your final creation. The mat or board should also be made from plastic or rubber so that you'll have an easier time removing the base from the board during the drying process.

2

With your hands or rolling pin, flatten out the clay until it's a little over ½in (1.2cm) thick.

3

Place the bowl or planter on top of the clay to mark the size and shape you'd like your base tray to be. With the detailing tool, cut into the clay and separate the part that you want to use for your base tray. Remove the excess clay and store it in a sealed container to use for future projects.

4

Take the spatula and, leaving a ¼in (5mm) border at the edge, press the spatula down into the clay about ¼in (5mm) deep. Continue scoring around the edge of the clay. You are marking out what will be the lip of the base tray. The lip of the tray can be as tall as you want but it shouldn't be shorter than ¼in (5mm).

5

With the single-wire end tool, scrape out the center clay. Remember that you want the base to be at least a ¼in (5mm) thick, so don't dig in too deep. Leave the excess clay to the side, just in case you need it later to reinforce an edge or to add somewhere.

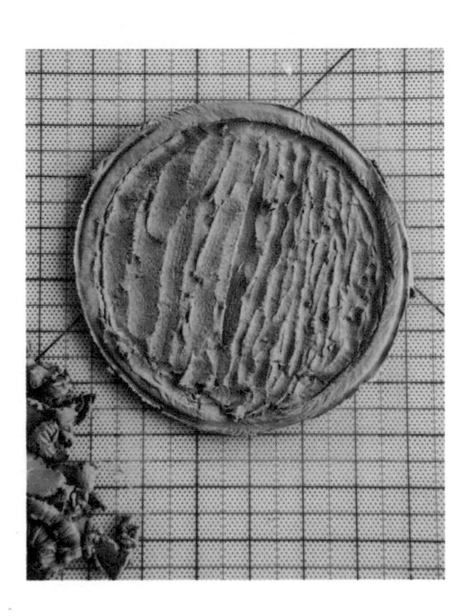

6

At this point you might want to spray the clay with water to make it more pliable. Use the spatula to help smooth out and build up the lip of your base tray. Then use your hands to smooth out the top surface of the base and the lip. Using your hands will give the tray more texture at the end but if you'd prefer a cleaner look, use a piece of foam board cut to the size of the inner portion of your tray. Press down evenly and then remove.

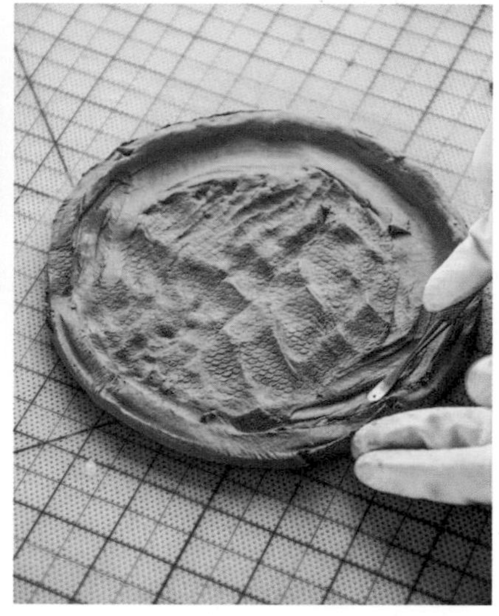

Side note I've chosen to make this tray quite simple but if you're looking to carve in any designs or cut shapes into the lip of the tray, this is the time to do so.

7

Keeping the base tray on the board, place your creation in a cool/dry spot. Let dry for 36 hours, then flip the tray over and let dry for another 36 hours.

8

Once dry, take the 120-grit sandpaper and sand the base tray to smooth out any imperfections. Clean away the dust and with your 220-grit, sand again to finish. Clean away the remaining dust.

9

Your base tray is complete! If you're happy with the raw color of it, place your potted plant on top but if you'd like to give it a little more pop, feel free to paint it with acrylic paint.

THE PROPAGATION STATION

When I first caught the houseplant bug and became someone that loved bringing plants into my home, the one thing that made my collection increase rapidly was my newfound love for propagation. I remember the day I was told that if I made a cut on my plant in a particular spot, I could take that cutting, place it in water, and watch as it developed roots, then place it in soil and grow a new plant. When I heard this my jaw dropped, my eyes widened, and, as my body recoiled a bit, I looked at that person and thought… "What are you, some kind of sorcerer?" I just couldn't believe that it was possible, but with knowledge of this magic, I gave it a try and was forever changed. In fact, I became obsessed. One day, I took my first cutting from a plant and then, by the end of the week, I had cuttings all over my home, in jars, old bottles, and, quite frankly, anything that could hold water. I found something so rewarding in propagating that it forced me to take more and more from my plants, from my family's plants, from my friends' plants, and, when I could, from my local nurseries. No, I'm not talking about stealing these cuttings. I definitely asked first. What I love about propagating is that you're able to take a cutting from a plant that you've spent so much time loving and caring for and gift it to someone that you also love and care for. I call propagation the gift that keeps on giving because when you gift a cutting to someone to propagate, it grows and develops and they can then take a cutting from it and gift it to someone else. So it goes from one home, to the next, to the next, and so forth.

PROJECT TIME
4 HOURS

In 2016 I had jumped from having just 80 plants to about 140 plants, all through the process of propagation. As you know, my goal is always to make a space feel lush but at the same time inspired, and when I moved into a new apartment with my wife, I really wanted to utilize the vertical space with a living wall. So I thought, why not create a wall full of cuttings in vessels. Not only would we be able to develop and grow these cuttings to expand our own personal plant family, but we would also have gifts for those that we care for whenever necessary. So, today I want to show you how to create your own propagation station using materials that you can find at your local hardware store and vessels you can find in your recycling. Throughout this book I mention a few ways to repurpose or upcycle materials that in most cases find their way into your garbage bin. Over the past few years, sparkling beverages have become a huge trend and in many cases, these drinks come in glass bottles. I figured what better way of creating a propagation station (while helping to reduce waste), than using these bottles. I drew out a design of what I wanted the station to look like and asked my friend and woodworker, Matt Norris, to help me construct a propagation holder using very few materials. After a few attempts, we finally got something we really were excited about. So, here are the steps for how to create your very own propagation station.

WHAT YOU WILL NEED

Ruler

Wood glue

Drill with ¼in (6.5mm) drill bit

1 x wooden board, ¼ x 3 x 48in (0.5 x 7.6 x 122cm)—we used oak but feel free to use any hardwood)

1 x wooden round dowel, ¼ x 48in (0.5 x 122cm) or 2 x brass rods, ¼ x 12in (0.5 x 30.5cm) if preferred

Lint-free rag

Spray can of clear-coat wood sealant

4 x quick-grip 6in (15cm) mini bar clamps

Miter box with saw

Cutting board

Sandpaper: coarse and 220 grit

Screws and anchors for wall mounting (optional)

4 x tall glass bottles

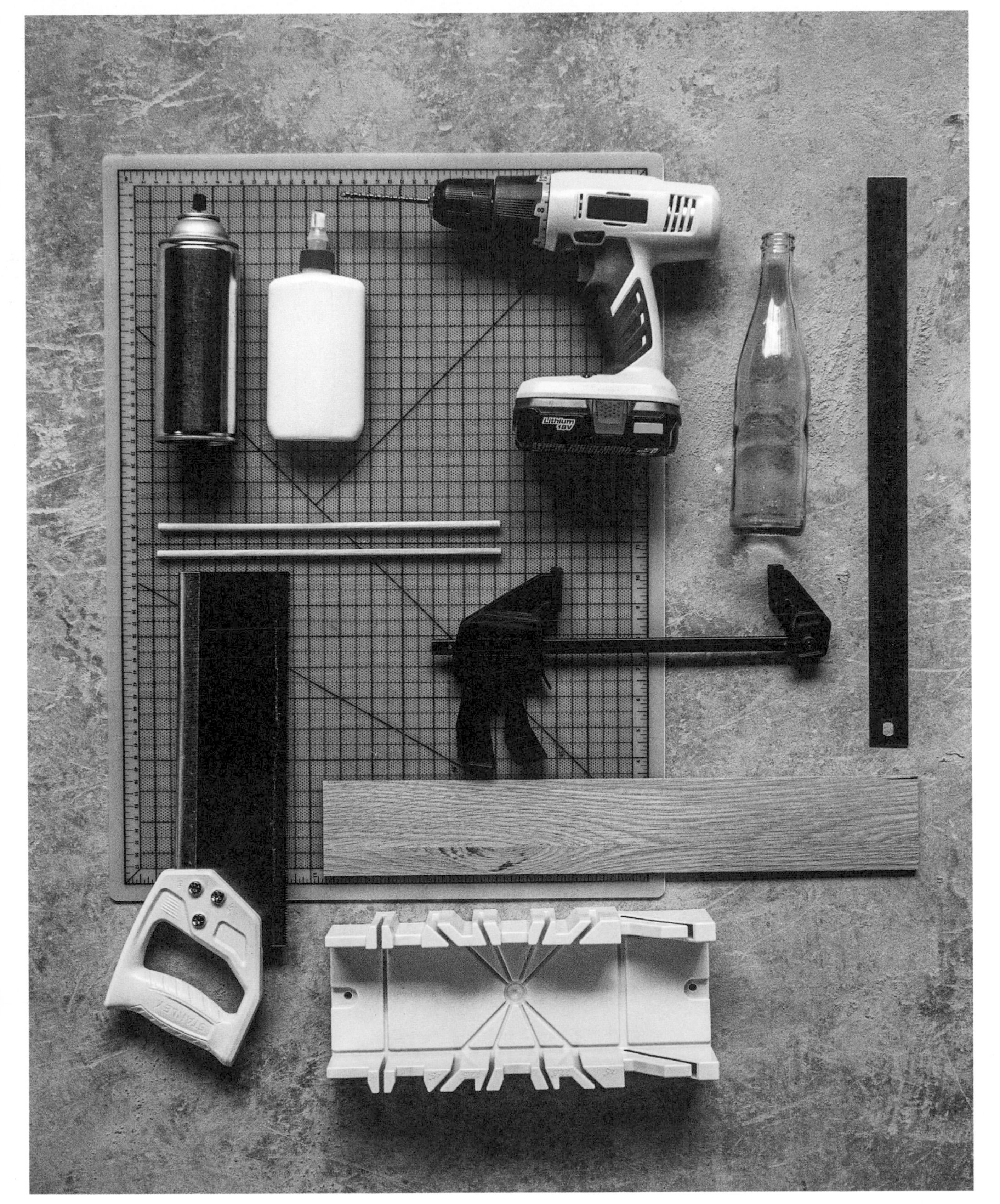
Lithium
18V

1

On a clean, sturdy surface, cut the board into the pieces needed for the project. Place the wood board on top of your work surface, and, using a ruler and pencil, measure and mark out the pieces you'll need to cut. First, mark out two pieces that are 12in (30.5cm) in length and then one of 11½in (29.2cm).

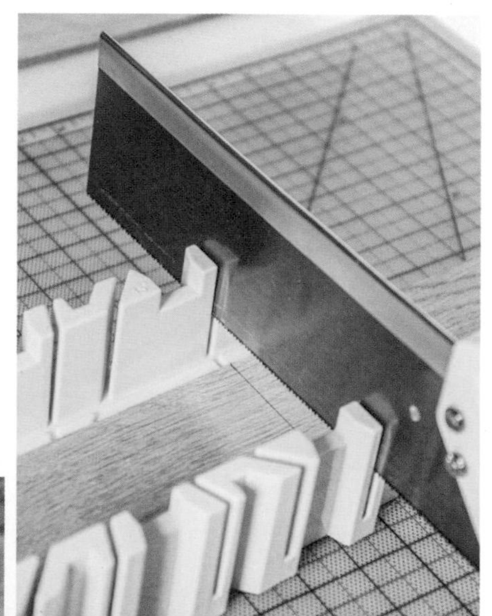

2

Place the miter box on the edge of your work surface and secure it to the table by clamping it down on both sides. Slide the wood board through the box and use an additional clamp to help hold it in place. Lining up the marks for where to make your cuts, take the saw and carefully make the cuts needed.

3

Now take the remaining piece of wood, mark out two trapezoids measuring 6in (15.25cm) long on one side and 3in (7.5cm) long on the opposite side. Take the board and line it up at the 45-degree-angle mark on the miter box. Take the saw and carefully make the cuts needed.

4

With the last remaining bit of wood, we want to create three small, right-angled triangles with 1½in (3.8cm) legs. To do this, measure out 3in (7.5cm) down the board and make your cut. You'll now have a 3 x 3in (7.5 x 7.5cm) square. Place that square in the miter box again and line it up at the 45-degree-angle cut and cut the board in half. You should now have two 3 x 3in (7.5 x 7.5cm) right-angled triangles.

Take both 3 x 3in (7.5 x 7.5cm) triangles and on the longer 4¼in (10.8cm) side, mark the halfway spot and cut them in half. This will leave you with four triangles, although you'll only be using three.

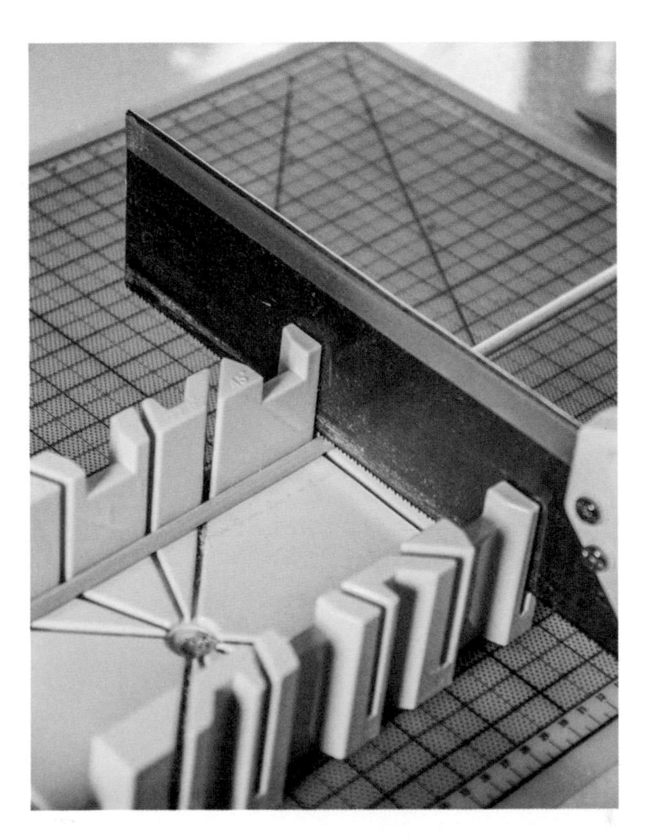

5

Take the wooden dowel and measure out two 12in (30.5cm) pieces. Make the cuts. You should now have all the pieces needed to make your propagation station.

6

Drill two ¼in (6.5mm) holes in each of the trapezoidal sides centered in the following locations: 2⅝in (6.5cm) from the back edge and ¾in (2cm) from the bottom, and the other at 2⅝in (6.5cm) from the back edge and 2¾in (7cm) from the bottom. Take care to make the holes perpendicular to the surface.

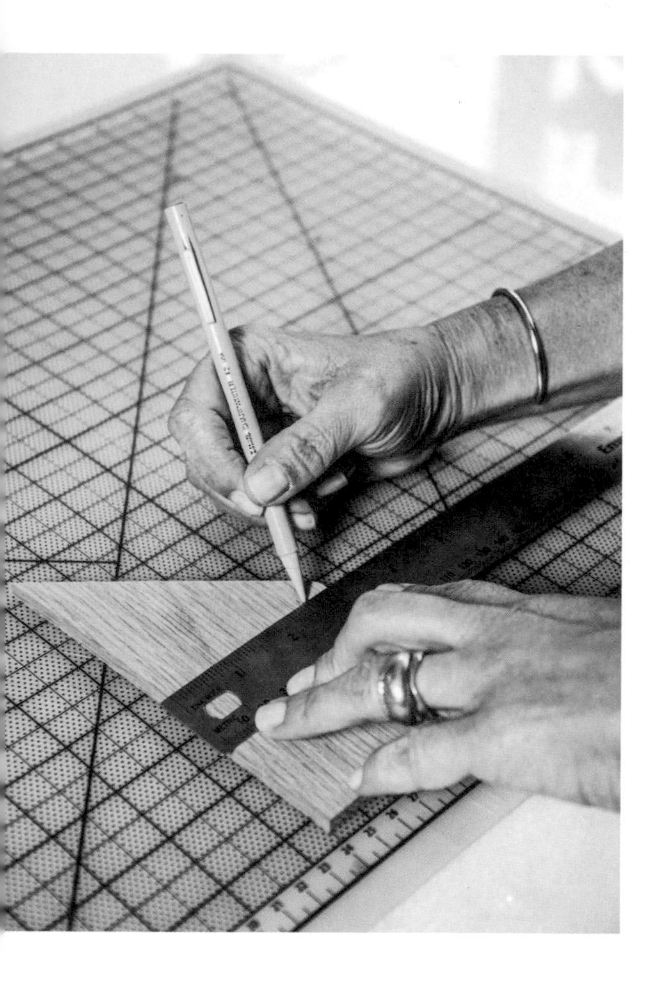

Tip Clamp the piece that you're drilling onto a scrap piece of wood to help prevent the bit from tearing out the back of the hole.

7

Now that you have the holes drilled correctly, make two little cuts straight down from the top edge to the sides of the top hole. This is a tricky cut, so maybe drill and practice on a piece of scrap first. Use sandpaper wrapped around a ruler to smooth the inside of the notch.

8

Take one of the 12in (30.5cm) pieces of board and measure out 1in (2.5cm) from the top and 2in (5cm) from the side. Do this on both sides of the wood. At these marks, drill ¼in (6.5mm) mounting holes. This is where you'll use screws to mount your propagation station.

9

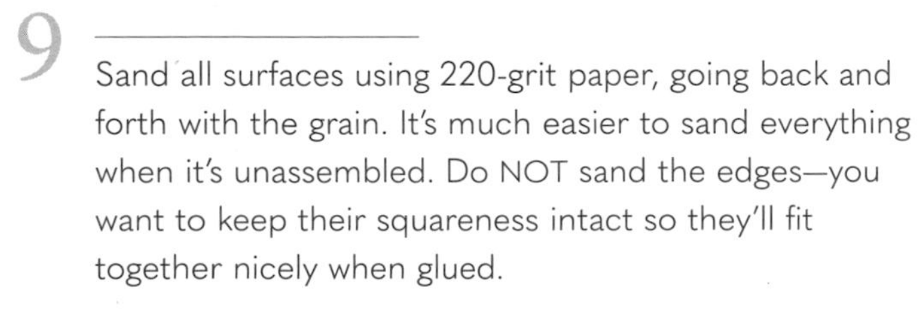

Sand all surfaces using 220-grit paper, going back and forth with the grain. It's much easier to sand everything when it's unassembled. Do NOT sand the edges—you want to keep their squareness intact so they'll fit together nicely when glued.

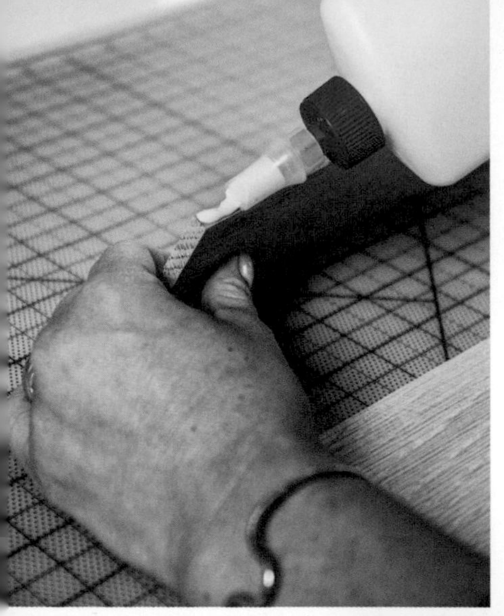

10

Apply a line of wood glue to the one long edge of a 12in (30.5cm) piece and then clamp the other 12in (30.5cm) piece to it, edge to edge, to form a 6 x 12in (15.2 x 30.5cm) piece. Initially clamp with moderate force, enough to hold the pieces together while still being able to slide them around so you can align them perfectly. Putting a clamp directly on the seam can help align the surfaces and keep the boards from folding. Once they are lined up, clamp them with full force and use a damp rag to wipe off any excess glue. Let the glue set for about an hour before unclamping.

11

Next, apply wood glue to the long edges of the two trapezoid side pieces. With the 6 x 12in (15.2 x 30.5cm) piece resting flat on your work surface (maybe cover this surface with wax paper or foil to prevent accidentally gluing your piece to it), clamp the two side pieces down onto the back. Without attempting to glue it at the same time, use the 11½in (29.2cm) bottom piece between the sides to help keep them square to the back.

Once in place, use a damp rag to wipe off any excess glue. When the sides are set (wait around 30 minutes), you can glue in the bottom piece, clamp it down to the back piece, and place another clamp to squeeze the sides together (place this clamp in line with the bottom piece so as not to stress the joints of the side pieces). When the bottom has set (around 30 minutes), glue in the little triangles, spacing them evenly across the width of the piece. Use glue sparingly for these, as there's no way to clamp them and it's not easy to remove excess glue.

12

Sand all the edges. Start with coarse paper and work your way up to 220-grit. "Break" any sharp edges by sanding them off: how rounded you make the edges is a matter of personal preference. Spray finish your piece with clear wood sealant outdoors only. Let it sit and dry according to the instructions on the can.

13

Once dry, take the wooden dowels and slide them through one of the bottom holes of your propagation station. Apply a little glue to both ends and slide the bottom dowel through to the other hole. Sand the ends flush with the surface of the sides. If using brass rods, make sure to remove any burr from the ends before inserting to prevent any tear-out of the holes. Cut and sand the ends of the removable top dowel/rod.

14

Securely mount the propagation station on your wall using screws and anchors if you wish. Fill the glass bottles with lukewarm water and place them in. Take cuttings from your plants or the plants of others, and place them in the bottles. Enjoy!

You can find Matt Norris on Instagram at @anatomatty

PRESERVED MOSS WALL ART

One of the questions I often get asked is if I think this re-emergence of houseplants will die away as it did back in the 1980s, after seeing so many bring plants indoors in the '70s. From my perspective it's never really died, and many were bringing plants indoors way before the 1970s. Maybe here in the United States, where I live, houseplants became trendy, fell out of style, and have now become trendy again. But in other places around the world, like Asia and Europe, blurring that line between interior and exterior has been ever-present. We as humans have lived among plants and nature since the dawn of time. We felt a connection with the outdoors then and still do today. As we've moved indoors, we've found ways to weave those outdoor, natural elements into our living spaces to stay connected. This art of living has a name and it's called biophilic design. If you're someone, like myself, who has been styling plants in your home, feels excited about hardwood floors or wooden furniture design, or has moved into new spaces because they had large windows, you are already seeing life through the lens of biophilic design and practicing its ethos. Biophilic design is all about finding ways to connect with the natural world by introducing those elements indoors. When you've heard me or others talk about how plants can make you feel more alive and creative, that's all due to the human connection with the outdoor world. We still, whether consciously or unconsciously, want to be at one with nature. That's why some people who work in large offices without a lot of natural light, stuck within a cubicle,

PROJECT TIME
45 MINUTES

will have a beach scene as their screen saver or a small faux plant on the side of their desk. These small reminders of the outdoor world trigger something in our brain to make us feel calmer and more productive. That's why you'll never hear me talk down to those who want to bring in faux plants. We don't all have the luxury of living in homes or working in offices with a lot of natural light, but surrounding yourself with faux plants, objects made from natural elements, or walls painted in earth tones can have the same effect on us as if we were outdoors.

I've been on a mission over the last nine years to find myself more connected, more at one with nature, and my home and workspaces have reflected that. I've created propagation walls which exist as living art but also preserved moss wall art that gives the look of life but doesn't demand the attention and care a living wall would. Whether it's alive or faux, if it looks as if it has come from the natural environment, it can help our minds, our hearts, and, of course, our lives. So while there are many projects in this book that will help you find creative ways to utilize live plants, I felt it was only right to include some projects that no matter what type of light or plant knowledge you had, you could create a space that blurred that line. Creating a moss wall is fun and easy and really helps to get your creative mind in motion. So here is what you'll need to create your own preserved moss wall.

STYLING INSPIRATION

Position moss wall art where you'll get the most from it. Take advantage of the fact that you're using preserved moss, which doesn't need water or light, and place the piece in an area where you can't keep live plants. Rooms like bathrooms or hallways work nicely. For me, giving my gallery wall some life and depth by adding a little greenery is always a nice touch.

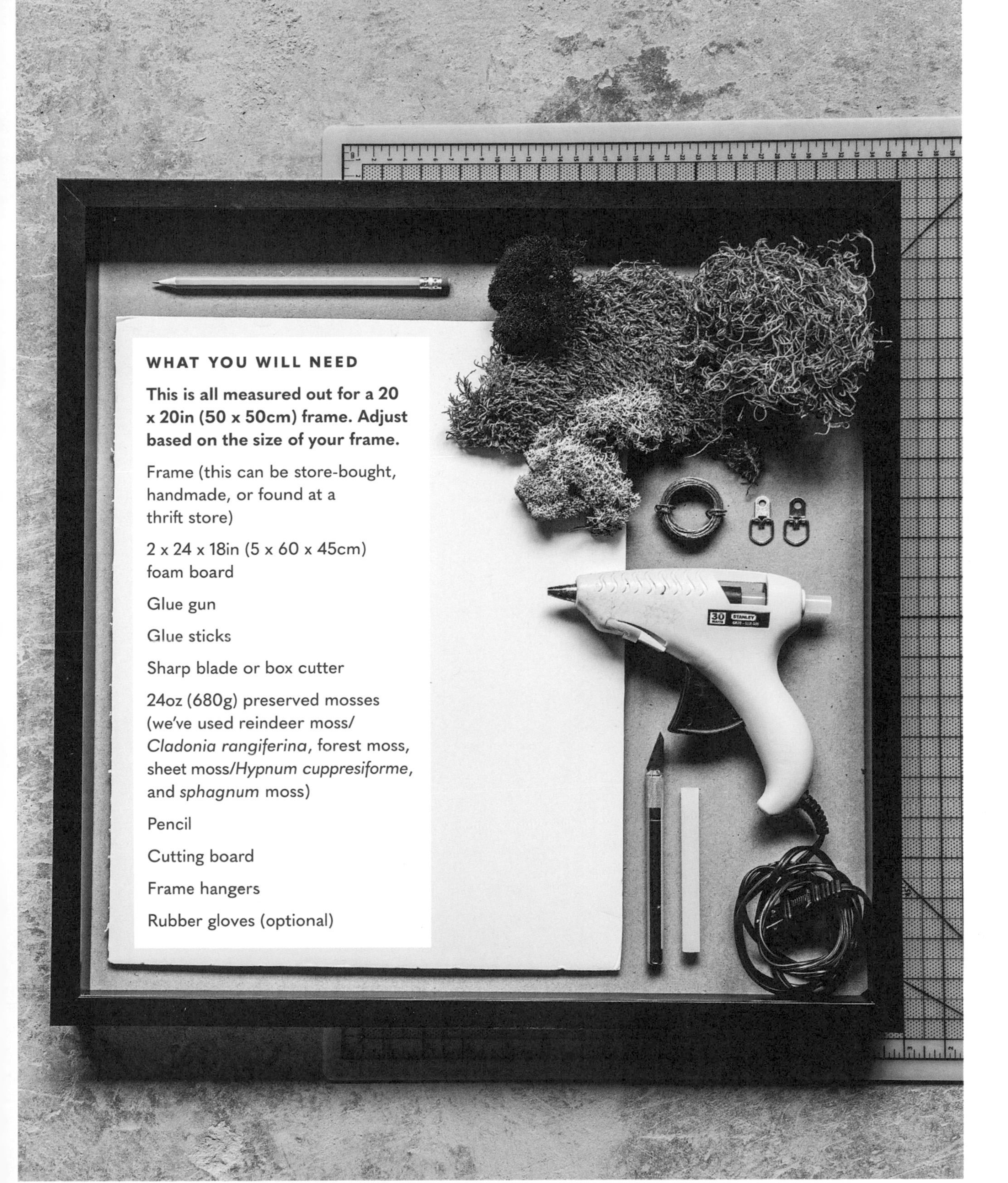

WHAT YOU WILL NEED

This is all measured out for a 20 x 20in (50 x 50cm) frame. Adjust based on the size of your frame.

Frame (this can be store-bought, handmade, or found at a thrift store)

2 x 24 x 18in (5 x 60 x 45cm) foam board

Glue gun

Glue sticks

Sharp blade or box cutter

24oz (680g) preserved mosses (we've used reindeer moss/*Cladonia rangiferina*, forest moss, sheet moss/*Hypnum cuppresiforme*, and *sphagnum* moss)

Pencil

Cutting board

Frame hangers

Rubber gloves (optional)

1

If purchased new, remove the frame from its packaging and then remove the backing of the frame so that you can take out the glass inside. Once the glass is safely removed, place the backing back in and secure it.

2

Place the cutting board on your work surface. Take the pencil and, on the foam board, draw out some shapes you want to use to give your art form and depth. These pieces of board will be stacked up to help create layers of mosses to give your art more of a natural terrain feel.

3

Using the sharp blade, cut out the shapes you drew on the board. Remove any tattered edges.

4

Place the foam board shapes inside the frame to get an idea of how you'd like the design to look. Once you've decided where you'd like them to go, take the glue gun and glue the foam boards into place.

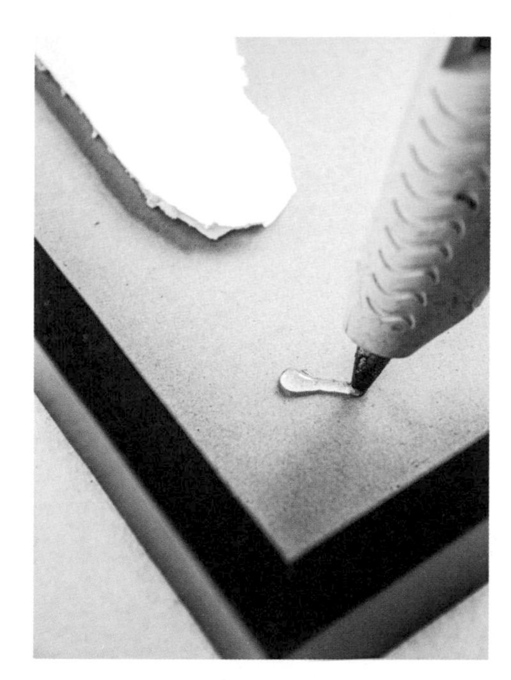

5

Here is where your artistic talent and creativity will be needed. It's time to place in the preserved moss. I wanted my art to have a flat valley winding through the middle, so I started by gluing in some sheet moss. For sheet moss, I feel it's best to place the glue on the back board of the frame and then apply the moss. If there are any pieces hanging over an edge that you don't like after gluing all of the sheet moss, use scissors to clean them up.

6

Start gluing the rest of the preserved mosses. Because these mosses aren't as flat as the sheet moss, I recommend putting the glue on the back of each moss and then pressing it down onto the board. If you feel like your fingers are touching too much glue, wearing a pair of rubber gloves will help. Again, be creative. There are dyed mosses that come in many different colors, so go wild!

7

Once the mosses all glued down, lift up the frame and flip over onto your work surface, and give it a gentle shake. You want to remove any pieces that aren't a part of the art but also see which parts need to be better secured with glue. Place back down and, again, cut away any moss hanging over the edge or places that you didn't intend.

8

Attach the hanging wire or picture hangers to the back and your moss art is ready to be displayed. Enjoy!

PLANT CHANDELIER

Did I ever tell you the story about the day I became obsessed with plants and knew I was forever changed? Are you sure? Well, let me freshen your memory. It was the moment I saw what I coined a "plant chandelier." Yes, I know that's not really a thing, but indulge me. In 2011, I was green when it came to plants, and not green meaning good, green meaning new. I had no idea how to care for a plant, honestly I could barely care for myself. Still, in that moment, as I walked into a greenhouse café, I knew I needed plants in my life. These plant chandeliers, as I called them, were in fact large staghorn ferns (*Platycerium bifurcatum*). These beautifully strange, robust orbs, floated above the café tables like traditional light fixtures, but alive and without luminosity. I knew from that moment I needed to start bringing plants into my life and style them in creative ways in my home. I understood that what would make a plant hanging over a table or in the center of a foyer powerful, was not just the type of plant selected to hang but also the vessel you place it in. When I saw those staghorns hanging above those tables back in 2011, it wasn't the fact that it was a plant I had never seen before hanging in a basket, it was the fact that it was hanging over a table. It's the unexpected, the non-traditional, that can make something so normal, feel so wild. I love that! And it's one of the rules I stick to the most when plant styling a space—always do the unexpected.

It was with that rule in mind that I decided to create a plant chandelier out of a model canoe. I mean, if you want unexpected, converting a 6ft (1.8m) canoe into a planter, then planting various like-minded tropical

PROJECT TIME
45 MINUTES

WHAT YOU'LL NEED

Wooden container e.g. a canoe or wine crate

Plants and soil

Spray can of clear-coat wood sealant

Heavy-duty plastic sheet (a heavy-duty trash bag will do)

Staple gun and staples

Paintbrush

Electric drill with ¼in (6.5mm) wood drill bit

4 x carabiners (size and color of your preference)

4 x eye hooks

2 x drywall anchors (if necessary)

Chain or rope to hang the chandelier

plants such as *Monstera deliciosa* and *Philodendron bipinnatifidum* (see *Combining Different Plants in One Pot* on p.142) and then hanging it over a dining table has to be a real example of the unexpected. It's in the space of the unexpected that you can create your statement piece. It's a conversation starter and lush canopy at the same time. I left that café back in 2011 inspired to use plants in a way that spoke to those around them. To bring life and art together seamlessly and also untraditionally. While you at home might not have the space to hang a canoe over your dining table, my hope is that this can inspire you to think outside of the box in finding a way to create your own plant chandelier.

Lithium
18V

1

It's important to first figure out where in your home you're going to be hanging your plant chandelier and how high or low you'd like it to hang. Once you've located that spot and measured the distance from the ceiling to where it will hang to, purchase rope, chain, or whatever you'd like to use to hang your chandelier with, a little longer than the length you need, to give yourself room for error. For these projects I decided to use chain because it added a little hardness to the softness of the wood and greenery.

2

Select a container for your plants. This container will make just as much of a statement as the plants, so be creative. Anything from a wine crate to a canoe will work. When making this decision, consider using hardwoods such as white or red oak that are more water-repellent, or woods like cedar that are bug-repellent. I used a salvaged wood container that was once used for industrial molds.

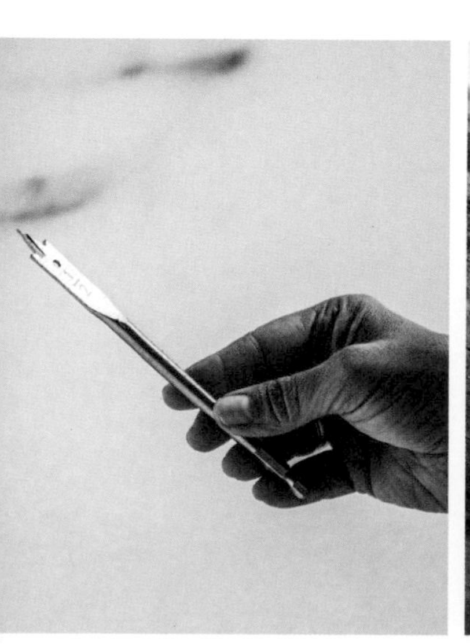

3

Take the drill and, with a wood drill bit, drill a hole in the bottom of the container to create drainage. You'll want to make sure your chandelier can properly drain to avoid overwatering and root rot.

4

Wipe away any sawdust or dirt with a dry rag. Open the can of wood sealant and with a paintbrush, evenly paint a coat to help seal the wood so that it can be well-preserved.

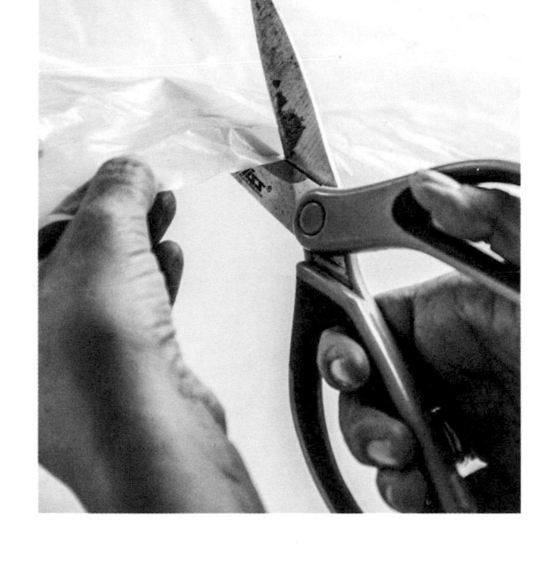

5

Measure the inside of your container and cut out a piece of heavy-duty plastic sheet to fit within that space.

6

Fold in the plastic sheet and, using the staple gun, staple along the top of your container to secure it. The plastic sheet is to better protect the wood from water damage and also helps to force all the runoff water through the drainage hole rather than the pores in the wood.

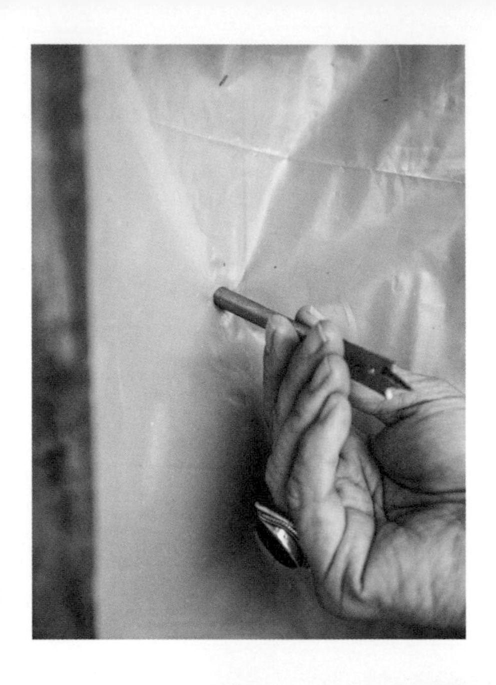

7

With the end of the drill bit, punch a hole through the plastic into the drainage hole.

8

Select the plants you want to place in your chandelier. If you're going to use different types of plants, make sure they require the same type of soil and moisture level. Once you know what types of plants you'd like to use, make sure to select plants with root balls that will fit in the base of your container with at least 2in (5cm) of space around them. We selected two *Aglaonema* 'Maria', an *Aglaonema* 'Silver Bay', and a *Philodendron bipinnatifidum*.

9

Pot your plants in your container. Remember to use a potting mix that works best for the plants you're using. I used a ratio of 80% potting mix and 20% perlite for these plants.

10

Screw the eye hooks into the sides of your container, making sure they are secure. Hook the carabiners around each eye hook and around the end of the chain or rope. Tighten the carabiners.

11

With the weight of the container, the soil, and the plants, making sure your plant chandelier is properly secured is very important. Please make sure to screw the second pair of eye hooks into studs or use proper drywall anchors. Hook the last two carabiners around the eye hooks in the ceiling and, finally, hook both ends of the chains to those carabiners. Tighten the carabiners and enjoy!

With your new plant chandelier, you're now ready to show off at your next dinner party. When it comes to watering your chandelier, to make it easier on yourself, water slowly, and place a small bowl or vessel below it to capture any runoff that comes out. You don't want to fill the container with too much water, causing it to become too heavy and possibly ripping it away from the ceiling.

HOW TO BUILD TERRARIUMS

To know me is to know that I'm a sucker for a good terrarium. There's something so magical and nostalgic about those small worlds of greenery encased in glass. Maybe it's because when I was a kid, I would often dream that somehow I'd be shrunken down like the kids in the movie *Honey, I Shrunk the Kids*. I wanted to fly on the back of a honeybee or jump through the mandibles of an ant. Terrariums create such beautiful living works of art. In their glass vessels, they are almost like sculptures that need to be displayed on a pedestal. I tend to find myself searching through thrift stores, consignment shops, and flea markets looking for vintage glass vessels to possibly build terrariums in. While the guts of the glass vessel hold the beauty and essence of the work, the glass vessel frames it, making it more presentable as a work of art. So, finding the right glass vessel can truly make the difference. I've seen some of the wildest terrariums created in old French press coffee makers, clear light bulbs, aquariums, lamps (hint hint), but it really does start with finding a vessel that makes a statement for the statement being made inside of the statement. I know that might be confusing but you get it.

The beauty of terrariums lies in how they create their own ecosystem. When making an enclosed terrarium, you never really have to think about watering it because it creates its own climate. Once closed, during the day, the moisture vapor inside the vessel condenses on the side of the glass and, as night comes, the moisture gathers in beads and races back down into the soil, keeping the humidity inside intact. There have been some cases where plants inside enclosed terrariums went years

PROJECT TIME
20–45 MINUTES (DEPENDING ON THE TYPE OF TERRARIUM AND SIZE OF VESSEL)

STYLING INSPIRATION

Place terrariums in areas that get bright indirect light but are also spots you can reach easily. Given that in most cases you'll need to keep the plants moist by misting them, don't place a terrarium high on bookshelves, behind larger plants, or where you'll need to pull a ladder out to get to it. Place your terrarium where you can show off its beauty front and center.

without being watered. So for someone in the plant community that loves the beauty of watching plants grow but doesn't have the time to water an additional group of plants every week, having an enclosed terrarium is perfect.

Because the soil in a terrarium remains pretty evenly moist throughout, I suggest that you only put in green life that thrives in moist soil. Ferns, alocasias, calatheas, to name a few, are great options in an enclosed vessel.

When you have a terrarium that opens, the self-regulating system of watering via condensation doesn't really apply here. So, like your other plant friends, you'll have to water them on their normal schedule, depending on the types of plants inside. Since some of the plants you'll be placing in your terrariums will be tropical plants, misting them at least once or twice a week helps.

When it comes to styling terrariums in your home, the first thing you always want to make sure of is that you have them in the right light. While some of your plants will do well in lower light, all of your plants would thrive in bright indirect light. The one thing you want to avoid when it comes to light entering your glass vessels is direct sun. This harsh light could heat up the interior of your enclosed terrariums, drying the soil and killing your plants. Once you've found the perfect light, treat them like an only child and display them so they can be the center of attention that they deserve to be. Clearly, I'm an only child. You've spent time creating this little world and you need to show it off. So if possible, hanging them is a wonderful idea, but I enjoy placing them on pedestals or plant stands, setting them in a space that makes it inviting for guests to come and take a look. When done right, a terrarium can be the statement piece in any home.

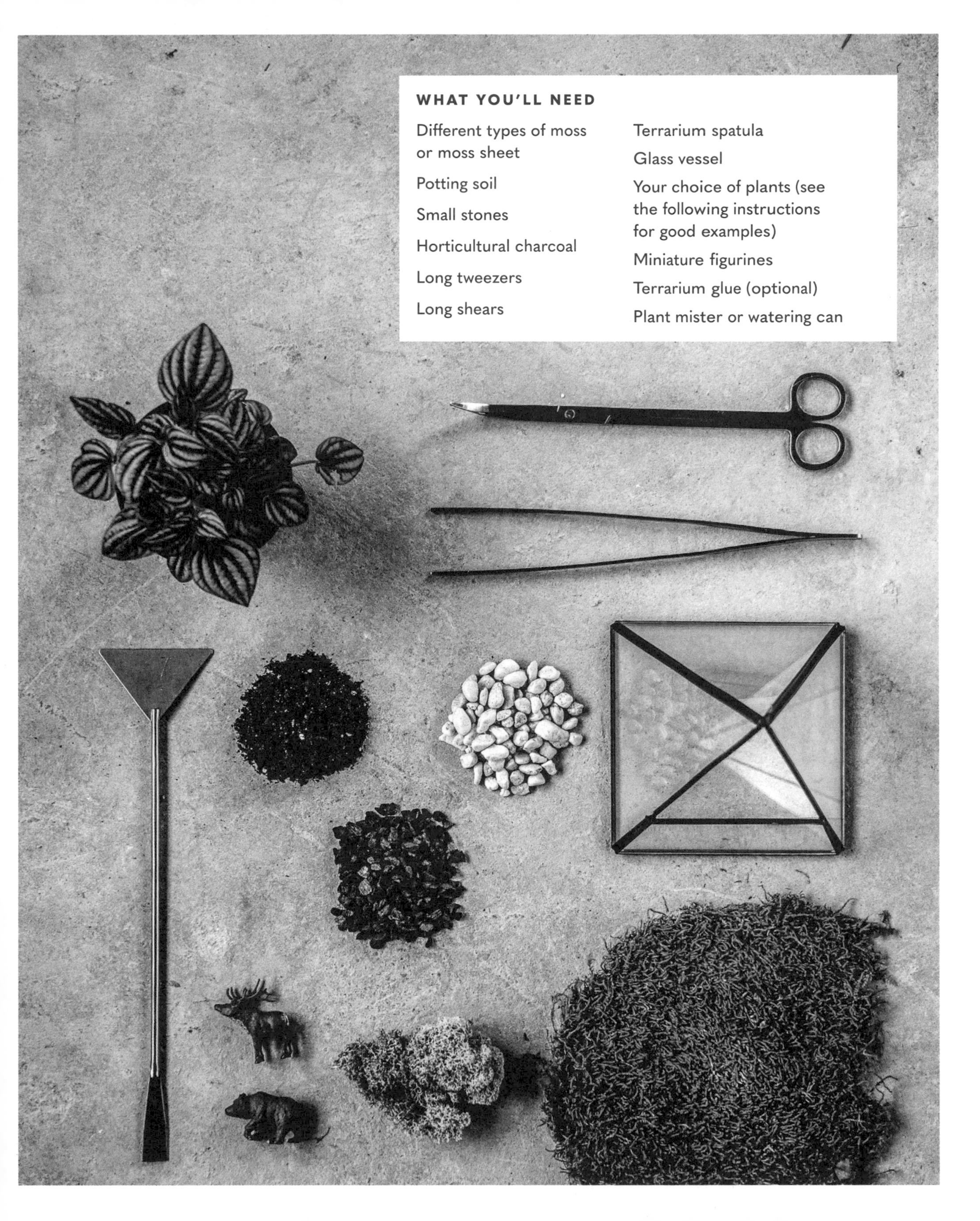

WHAT YOU'LL NEED

- Different types of moss or moss sheet
- Potting soil
- Small stones
- Horticultural charcoal
- Long tweezers
- Long shears
- Terrarium spatula
- Glass vessel
- Your choice of plants (see the following instructions for good examples)
- Miniature figurines
- Terrarium glue (optional)
- Plant mister or watering can

Air plant terrarium

Air plants are some of the best plants to style in a terrarium vessel because they need a lot of humidity. They are also one of the easiest plants to create a terrarium with because one, you're not potting the plant in the vessel, so you can take it out when it's time to water it, and, two, because you don't have to worry about soil, you have so many more options for where to style it. As long as you have the appropriate light, air plant terrariums can go anywhere. One of my favorite things to do is to treat them like ornaments and place them in my larger trees around the house. Why is it that the Christmas tree is the only tree that gets to be dressed up? My fiddle-leaf wants a little love too.

When it comes to making an air plant terrarium, here is how you should go about it.

1

Find the glass vessel that fits the space you're looking to place it in. Dress the bottom of the vessel. Because you're not using soil here, you can get creative. I've placed small stones at the bottom of my vessel, but feel free to get imaginative. Go WILD! If a child is making one with you, placing Lego pieces at the bottom could look cool.

2

Add your moss. Again, this will be for decoration, so style it the way you see fit. I've added a bit here to make it feel more bush-like.

3

Add your unique touch. My goal for this one was to be a little subtle and have a bit of a beachy vibe, so I found the perfect piece of driftwood to lay on top of the stones.

4

Lastly, place in your air plants. Again, depending on the size of your vessel, the amount you place in will vary. For me, as with every other moment in styling, you want to get the balance right. I like it when the air plants sit nice and cozy in the vessel while still crawling out to break the hard lines of the glass. Here, I used a sky plant (*Tillandsia ionantha*) and a *Tillandsia bulbosa* to protrude from my terrarium. The difference in their size and shape brings the wildness I love when creating a terrarium like this.

Glass lantern

One can only have need of so many candles, but as you read in *Vibe Setters* (p.156), candles are an important part of making a space feel warmer and more alive. So, with more people bringing candles into their homes, there have been many great glass lanterns designed so you can safely and stylishly place them throughout your space. It was in these great designs that I saw more than a vessel for old-school illumination: I saw another opportunity to create a terrarium. My saying is, if you can't find a place to put another plant, you haven't looked hard enough. What makes a lantern the perfect vessel to create a terrarium in is the fact that lanterns usually come with a handle so you can carry them from one dark room to the next. Having handles also makes them the perfect vessels for hanging in the home. It gets them off the floor or the surface of a table and allows them to float at eye level so their beauty can be more easily admired. Another reason why lanterns are a great option is because a lot of them come with glass doors that will allow you to display your terrarium with the door open or closed. Lanterns are never fully enclosed, so you'll have to water your lantern terrarium according to the care needed by the plants you place in them.

For my lantern, I choose a maidenhair fern (*Adiantum raddianum*) and a small *Alocasia* 'Mandalay'. With these both requiring the same care needs, placing them in the same container is possible (see p.142 for *Combining Plants in One Pot*). The dappled/filtered light of my eastern-facing window gives them the perfect light to thrive and because they require their soil to be evenly moist, I water at least twice a week using a long-spout watering can and mist every other day.

1

First, clean the glass of the vessel because once your plants are inside, it'll be difficult to remove any smudges.

2

Next, place a small layer of stones or rocks at the base to create a nice buffer zone. This area is to hold runoff moisture and to keep your plants' roots from sitting in water. Use the spatula to even out the stones.

Place a layer of horticultural charcoal on top of the stones to increase a buffer but also to help reduce impurities that can build up.

3

Place in the soil. I wanted to build a little hill for my terrarium, so I placed more soil to the back of the vessel. Once the soil is set, decide where your plants will go and then remove them from their nursery pots, gently break up the soil to loosen the roots, and then place them inside the terrarium. Cover the roots with a bit more soil and use the spatula again to even it out.

4

Measure the sheet moss and cut out the appropriate pieces to fit around your plants. With the sheet moss, it can play as grass in a field, so I like to use it in a way that creates a small valley. Once that's laid, I'll place in other pieces of moss.

5

Make it feel a bit more like a small world by adding other stones, pieces of wood that feel like fallen trees, and, lastly, a little figurine. Here I've placed in a small elk to make it more WILD!

Finally, display it in an area of your home or office that gets the appropriate amount of light for the plants you placed inside to thrive. And voila!

Wardian case

If you've read my book, *Wild Interiors*, you'll know of my love for historic conservatories. I've visited some of the most amazing green-filled structures around the world. From the Kew Gardens Conservatory in London, UK, to the Conservatory of Flowers in San Francisco, CA. The true magic of these glass and metal green spaces is that they house some of the rarest plants from around the globe. My obsession with greenhouses was probably created by Nathaniel Bagshaw Ward, the inventor of the Wardian case. The Wardian case is the original terrarium, and back then they were designed with the intention of easily transporting plant life from one part of the world to another. Today, they aren't seen as much unless you're some sort of plant hoarder (yes, you, the one reading this, I'm talking to you!). And given that I'm what you might call an habitual plant hoarder, having a Wardian case in my home is a must-have.

To have a Wardian case is to have your own little conservatory inside your home. I mean, they are shaped like miniature greenhouses, made from glass and metal, and designed in that vintage Victorian style. What's not to love?! These vessels provide you with a space in which to get truly creative. You can bring in more plants, more décor, and, of course, more WILD! The Wardian case won't have enough depth for you to plant directly inside it, so you'll need to have your greenery potted and displayed inside. I chose a *Phlebodium aureum* 'Blue Star', an asparagus fern (*Asparagus setaceus*), a *Peperomia albovittata*, a calathea Freddie (*C. concinna*), and a Japanese holly fern (*Cyrtomium falcatum*) because not only does their foliage work well together but they also all require the same type of care.

1

Unlike the other terrariums in this section, since you're planting directly in a pot, creating a buffer zone at the bottom isn't necessary. Just make sure to use a pot with a drainage hole.

Place in your soil. Make sure that a third of your pot is filled with soil.

2

Remove your plants from their nursery pots, loosen the soil and their roots, and gently place them inside, arranging them in a way that will allow them all to get exposure to light while also creating a beautiful design.

3

Add the top layer of soil, leaving at least 1in (2.5cm) at the top of the pot free.

4

Water your plants. Since your Wardian case will most likely be fully enclosed, you probably won't have to water it again. But if you don't notice condensation building up on the sides of the glass, that means you'll need to mist it every few days and probably water here and there to keep the soil evenly moist.

Display and enjoy!

Beaker glass

Plant care is a part of biology and biology, well, is a part of science. So, using beakers to create terrariums or propagate your plants just makes sense. These vessels can be found in many thrift stores, vintage stores, and, of course, online. I like repurposing them to help expand the indoor oasis I have created in my home. The shape of the glassware sets them apart and makes them easily recognizable. Beakers tend to be a bit more delicate than other glass vessels, so it's important to place them in areas of your home that aren't in the path of foot traffic or paw traffic (our cats have knocked over a few plant friends over the years).

I decided to create a little beaker terrarium that would sprout its vines out of the top and, as it grows, cascade down the glass. Give the size of the beaker, I wanted to go with a plant that had small, delicate foliage that wouldn't crowd the interior of the vessel too much, but would still grow perfectly in this setting. So going with a creeping blue *Pilea glauca* felt like the perfect choice.

1

Again, always start by cleaning the glass of the vessel because once your plants are inside, it'll be difficult to remove any smudges.

Next, gently place a small layer of stones or rocks at the base to create a nice buffer zone. This area is to hold runoff moisture and to keep your plants' roots from sitting in water. Depending on the size of the beaker, you could probably just move the vessel from side to side to even out the stones.

2

Horticultural charcoal isn't necessary here but feel free to add a layer if you wish.

Fill one-third of the vessel with soil.

3

Remove your plant from its pot, loosen the soil and its roots, and gently place inside. Use tweezers to help maneuver it down into the vessel.

4

Add a bit more soil to cover the roots. Use a small spatula to compact the topsoil a bit.

5

Give your plant a drink. When dealing with an opening that's small like this, to make sure you give the right amount of water and don't flood the bottom of the vessel, I suggest taking a wooden dowel or chopstick, placing it into the vessel, and, with a bottle of water, slowly pouring water down the side of the dowel so that you can accurately control the amount of water your plant gets.

JUNGLE BY NUMBERS

There are many creative ways to blur the line between indoors and outdoors, so that your home takes on a life of its own—your own personal paradise. Over the past few years, as houseplants and biophilic design have become more popular, weaving greenery into the indoor space has become just as common as seeing greenery outdoors. Bringing in plant life is one of the main ways we achieve this look, but sometimes a living plant is just the first layer. Many have taken it even further. We've seen it in fashion, home décor, art, and books. Plant prints, living walls, live-edge wooden products, etc. have been popping up all over the place. And when working in biophilic design, that's a large part of it. It doesn't always have to be a living plant that can bring that element of warmth and calm when you see it or when you're around it. It can be a faux plant or an image of a jungle. These images, or a wood-slab dining table, can remind you of that vacation you took or have been meaning to take, and fill you with that same excitement and peace. So knowing that not everyone has the knowhow and, sometimes, the desire to care for plants, I figured why not give those in the green-loving community a way to add that bit of life to their homes without worrying about tending to actual life.

I reached out to my good friend Drury Bynum, a painter, director, and just an all-round great artist, and asked him to help me design a jungle-themed mural that plant lovers could create at home themselves. I felt the best way to make it more achievable, regardless of artistic prowess, was for it to be a paint by numbers process. To make it even

PROJECT TIME
3 DAYS (DEPENDING ON SCALE)

WHAT YOU'LL NEED

Brushes: size 16 Flat, 10 Flat & 6 Flat

Vine charcoal

Painter's tape

Permanent marker

10 x quart-size (1-liter) containers with lids

1 x large 4-quart (4-liter) mixing container

4K projector

Drop cloth (size depends on how wide your wall is)

Bucket for water to clean brushes

Rags

Stir sticks

3 x 1-gallon (5-liter) cans of paint: white, mid-tone, and dark

Line art, see p.236–237 or download PDF of artwork here: thingsbyhc.com/junglebynumbers

easier, I challenged Drury to make it a limited palette. And by limited, I mean basically monochromatic. The idea was to select three base colors. One of those colors would be white and the other two would be your main color and a darker shade of that color. For example, let's say you wanted the main color of your jungle mural to be gray, you'd select a medium gray hue, a dark gray hue, and a white. Now that you have your main colors, the goal is to create a gradation between the three so that you have a total of ten colors. Oh yeah, you'll be mixing some colors here so you might want to get your overalls ready! Drury and I discussed the types of plants that should be in the mural and I thought the larger the foliage, the better. Simple shapes like *Monstera* leaves or fiddle-leaf fig (*Ficus lyrata*) leaves. But of course Drury, being the artist he is, went full WILD! When he shared with me the sketch that he wanted to use for the mural, my jaw dropped and I'm sure yours will as well. This is why I'm so excited to share the project here with all of you.

What I love about this project is how challenging and involved it is. But not in a way that will make you toss your paint out the window and give up. It's challenging in that way that makes you think deeper about each move you make, the brushstrokes you choose, etc. It's also a great project to do with a group. If you find yourselves with an open weekend and looking for a fun activity, this is the perfect project to do with the family or your friends. Given that it's all numbered, each family member can take a number or two and everyone can create together. So here are the steps for how to create your own Jungle by Numbers.

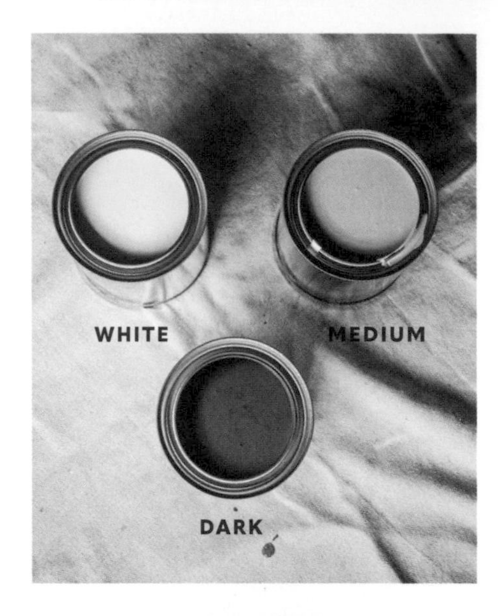

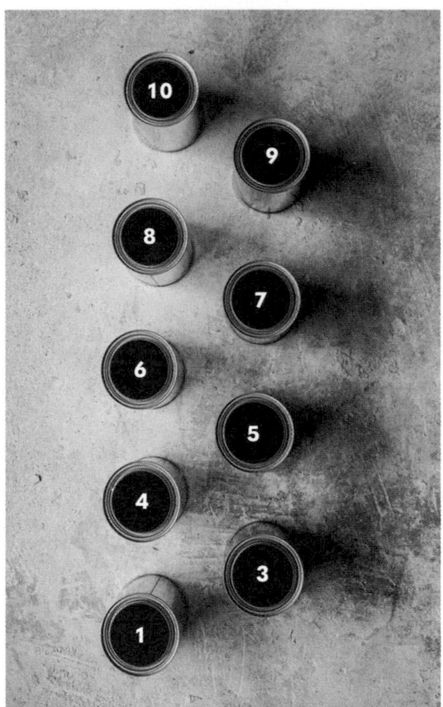

1

Pick your colors to create your jungle. For the mural here, we decided to go with a green color scheme, so we selected a medium-hue green, a dark green, and, of course, white. These will be our three main tones. Please don't feel as if you need to go green with your creation. You can use any color scheme you prefer; just make sure you have a range of three distinctive tones from light to dark.

2

It's now time to prepare your surface. Make sure your wall is paint-ready by wiping it down to remove any dust or debris. If there are any surface imperfections you need to clean up, fill those holes with spackle (polyfilla), sand, and once ready, prime the wall.

Note It's best to paint on top of white or neutral tones.

3

Before you start mixing the paint, place a bit of tape on the side of each 10 quart-size (1-liter) container and using your permanent marker, label them from 1–10. Be sure to label the lids as well. Arrange your containers into two side-by-side rows. For now, we're leaving out container #2, as it will be mixed at the end. You will end up with two rows: the first row includes containers #1, #4, #6, #8, and #10, the second contains #3, #5, #7, and #9. Remember, container #2 will be mixed later.

4

Create your base colors. These are the three main colors you chose. Carefully open your 1-gallon (5-liter) cans and stir the paint using a stir stick. Once ready, fill containers #1, #6, and #10 with your white, mid-tone, and dark colors, respectively.

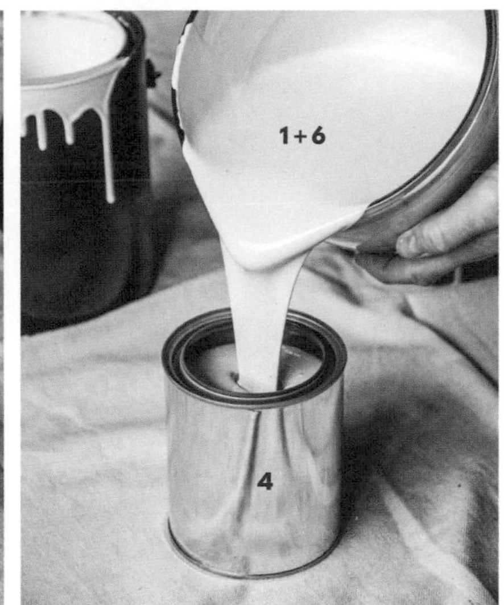

5

Create your mixed colors. From your 1-gallon (5-liter) cans, mix the following in your 4-quart (4-liter) mixing container: 1 quart (1 liter) each of color #1 and color #6. Mix thoroughly and fill container #4. Then, from container #4, fill containers #3 and #5 each halfway. Refill container #4 with the remaining 1 quart (1 liter) from the mixing container.

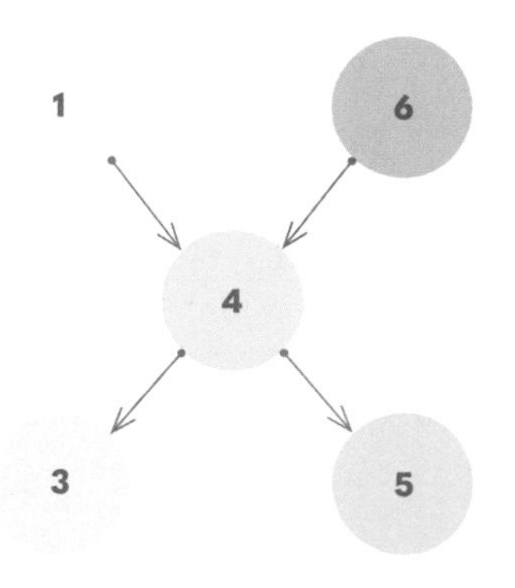

6

Fill the remainder of container #3 with color #1. Fill the remainder of container #5 with color #6. Using a stir stick, mix the colors thoroughly.

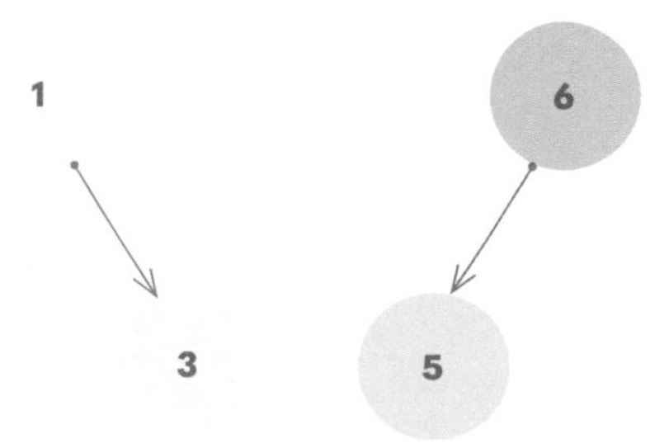

7

Clean the 4-quart (4-liter) mixing container. Refill container #1 and container #6. Again, from the 1-gallon (5-liter) cans, mix the following in the 4-quart (4-liter) mixing container: 1 quart (1 liter) each of color #6 and color #10. Mix thoroughly and fill container #8. Then, from container #8, fill containers #7 and #9 each halfway. Refill container #8 with the remaining 1 quart (1 liter) in the mixing container.

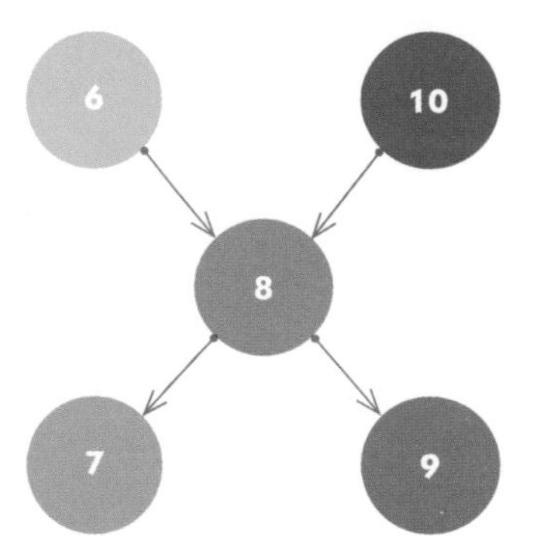

8

Fill the remainder of container #7 with color #6. Fill the remainder of container #9 with color #10. Using a stir stick, mix the colors thoroughly.

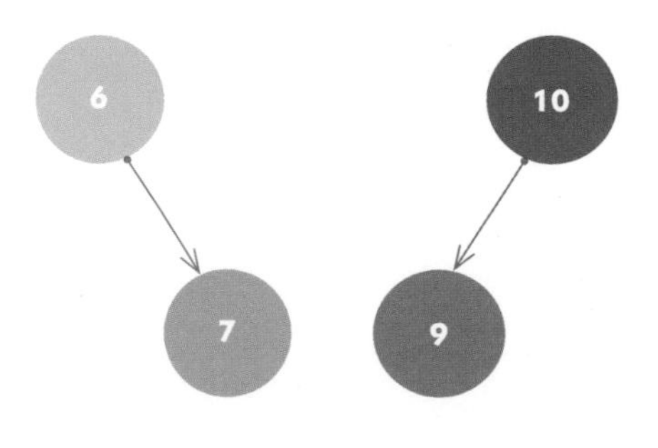

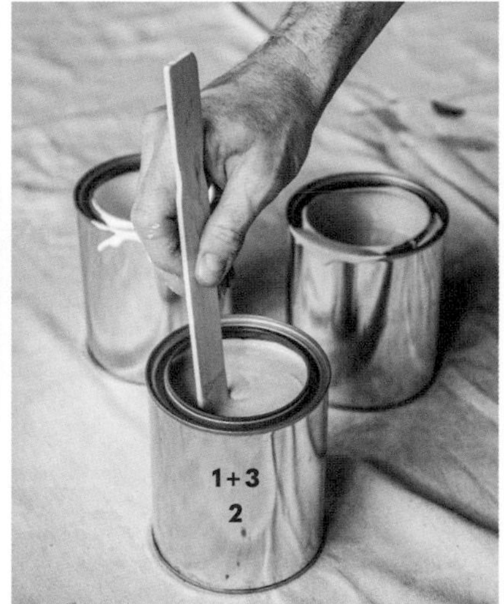

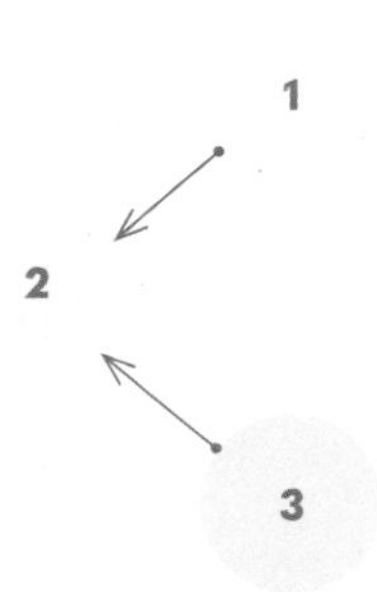

9

Refill container #6 and container #10. Since color #1 is nearly pure white and color #3 is a strong green (in the color choices for our mural), we needed something closer to white than green for color #2. This is why we set aside container #2. Fill 90% of container #2 with color #1. Fill the remaining 10% of container #2 with color #3. Using a stir stick, mix the colors thoroughly. Place the lids on top of each container.

Note Color #2 is your cloud color and color #1 is the highlight on the clouds.

Now the fun really begins! It's time to project the line art. Use your projector to cast the image on the wall. We ask that you use a 4K projector to give yourself the clearest lines possible. For best results, make the room as dark as you can. Using the vine charcoal, follow the line art as closely as possible to draw out the mural. Make sure that all shapes are closed shapes that contain a single color assignment. Draw the numbers in as you go.

Note In the line art, every single shape is assigned a number. There may be places where you feel that level of information isn't necessary to include in your drawing. For instance, in this section above, the water and shoreline behind the plant are labeled between every spike. You may feel recording every number is unnecessary if you can easily visualize these foreground/background layers.

11

Paint in the colors corresponding to the numbers on the cans. Try to cover the charcoal as you go. You'll find that the charcoal will mix with the paint, but it shouldn't affect the colors enough to be noticeable. Continue until the mural is complete!

Note The darker the color, the more likely it is that more than one coat will be needed. You may prefer more coats, as the visible brushwork creates texture in the mural. However, to get the best tonal range, two coats are recommended.

You can find Drury Bynum on Instagram at @drurybynum

THE AIR PLANT WREATH

'Tis the season to make a wreath! OK, well not your traditional wreath. For the most part wreaths have only been displayed during the fall and winter seasons as a religious ornament and, if you go way back, as a symbol for victory in the days of the Roman Empire. Over the last few decades, I believe wreath design has moved into an area that's more about giving your front door or the space above your mantle a little extra flair. When I was a kid, my mother and I lived in small apartments and never really felt it necessary to place a wreath on the door, but now that I'm older, I see the wreath as being as important as the welcome mat that sits below your door. A really nice wreath on the front of your home is like a beautiful bow on a gift, a pearl necklace caressing a neck, or the perfect planter in which to present your favorite plant. I knew I'd find a way to tie this back to plants!

And speaking of plants, while most wreaths have mainly been constructed with live florals that will die over time or dried florals that remain intact, not many were made with living plants that would remain alive while hanging in or on the home. I mean, it makes sense. How would you attach the planter, holding the soil and roots of your plant, to the wreath? I would assume that would be difficult, not to mention unsightly, especially for something that is supposed to be about style. So in trying to become more of a wreath family, my wife challenged me to create a wreath that we could hang on our door or above our mantle during the warmer months of the year. Not sure if you know this about me or not but I kind of have a thing for

PROJECT TIME
1 HOUR

houseplants. So I thought, this is the perfect opportunity to bring in more greenery in a way that was artfully done and not have my wife yell at me for bringing in even more plants. Yeah, I told you I was a plant hoarder.

With the spring and summer in mind, and wanting to make a wreath with plants that could continue to live while adorning it, I thought there's no better family of plants than air plants (*Tillandsia*). Air plants are so great for tucking into smaller spots in your home that you're looking to freshen up with a little life. And given that they don't grow in soil, are fairly lightweight, and love humidity, they make the perfect main feature for a spring/summer wreath. Of course, like all of your houseplants, you'll want to make sure you are still able to provide them with the proper care, so please consider the light and humidity in the area of the home where you want to place them first. I live in an area that is very humid during the warm months of the year, so if I placed my air plant wreath on my front door, I wouldn't have to water or mist it as much. But this isn't the same for every home in every climate. (For more information on how to care for air plants, see p.198.) So, if you're looking to dress up your front door or mantle during the spring and summer, here are my tips for creating the perfect air plant wreath.

WHAT YOU'LL NEED

19in (50cm) floral hoop

Wire cutters

22-gauge floral wire

Shears

Floral tape

Rope

Spray bottle

Glue gun (optional)

Wreath hook, nail, or command strips to hang the wreath

Air plants used

Tillandsia stricta

Tillandsia capitata

Tillandsia seleriana

Florals used

3 *Magnolia* stems

4 dried pampas grass (*Cortaderia selloana*) stems

20 dried bunny tails (*Lagurus ovatus*)

2 dried billy balls (*Craspedia glauca*)

1

Lay the hoop down on a clean surface and place your air plants and florals where you imagine them going. This will give you a good idea of how everything should be arranged and will look once complete. I know I don't need to remind you, but be creative and have fun! I've decided to use a large *Tillandsia stricta* to be the main feature of my wreath because of how structural and bold this air plant is.

2

Position your first layer of florals. For this wreath, it'll be the pampas grass because I'll want that to be the backdrop for the entire layout. Using the wire and wire cutters, cut a 3in (7.5cm) piece of wire and tightly wrap it around the stems of the pampas grass and the floral hoop to secure them together.

3

Once the pampas grass is secured to the wreath, place your main air plant at the far end of the wreath, but let a little pampas grass lead the way. Cut a 12in (30cm) piece of rope and carefully tie it around the bottom of the air plant and the hoop. Weave the rope through the outer leaves at the base of the air plant. These leaves will be drier and hold the rope better.

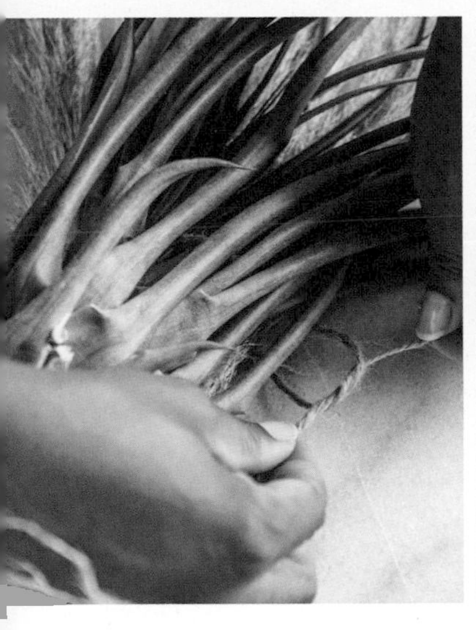

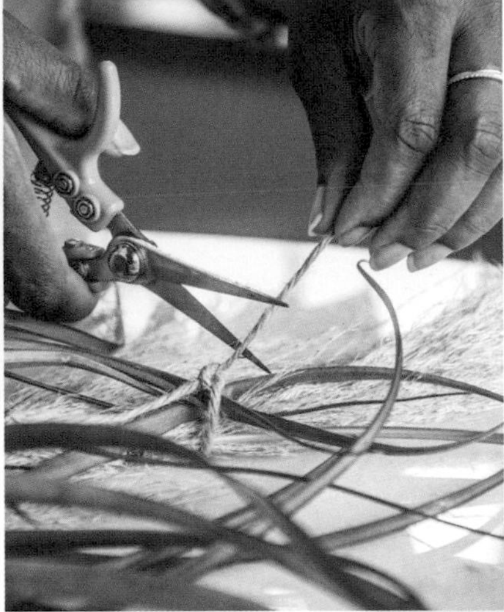

4

Then cut a 6in (15cm) piece of rope and carefully weave it through the middle section of the air plant, loosely tying a loop around the last few leaves and knotting the loop. The leaves should be resting in the knotted loop, not tightly secured. Cut off any excess rope.

5

Again, using the wire and wire cutters, cut a 3in (7.5cm) piece of wire and tightly wrap it around the stems of the *Magnolia* and the hoop. Repeat this step with any additional *Magnolia* stems.

6

Take the bunny tails and in one hand, gather them as if making a bouquet. Make it so that the bouquet is heavier at the top and thins out at the bottom. Take the floral tape and pull off a 3in (7.5cm) piece and, while wrapping it around the stems of your bunny tails, fold and twist the tape to make it become adhesive.

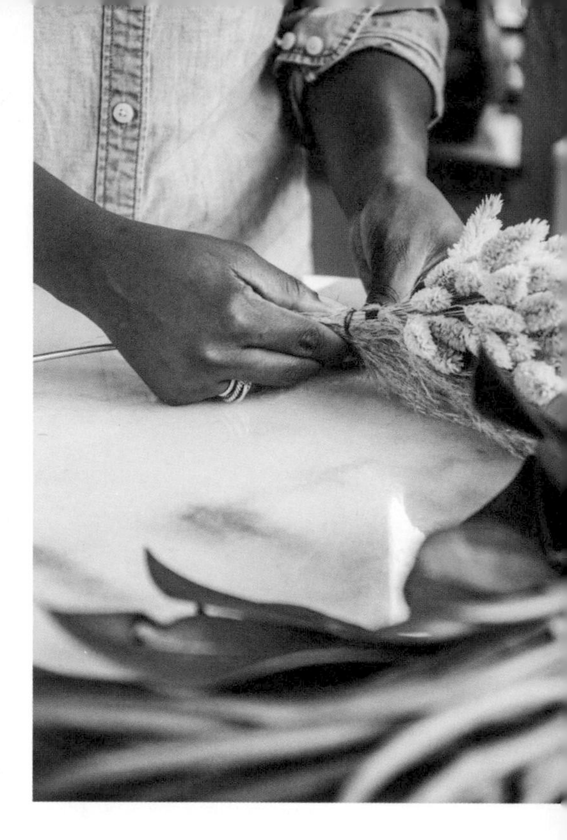

7

Once taped together, take the bunny tail bouquet and place it at the end of the floral design, where your pampas grass starts. Using the wire and wire cutters, cut a 3in (7.5cm) piece of wire and tightly wrap it around the stems of the bunny tails and the hoop.

8

Cut 18in (45cm) of rope and tightly wrap that around the base of the bunny tail and pampas grass stems and your hoop. Once you get to the last bit of rope, loop it through itself to secure it. This will help give your wreath a cleaner, more finished look. Continue cleaning it up by cutting away any excess stems, wire, or rope that are sticking out.

9

Add your final touches. For this wreath, I used two billy balls to add a pop of color. Take these stems and just work them through the other florals. You can use floral tape, wire, or even a glue gun to secure them. Your air plant wreath is complete!

10

With the spray bottle, spray the air plants with lukewarm water so that they get the moisture they need to start with. If you're placing the wreath outside, and you live in a dry climate, you'll need to spray your air plants at least once a day.

Securely hang from your door with a wreath hook, nail, or even command strips. Enjoy!

WILD HACKS

Over the past few years I've seen this green-loving community grow vastly and it's been one of the best feelings ever. In an environment where plant lovers like myself were few and far between, it's been a beautiful thing to watch as so many like-minded individuals have come together and bonded over a common passion for bringing the outdoors inside. In this passion, we have realized that we're working with life and like all living things, life is fragile. It's in that fragility that we try our very best to ensure our plants have the best lives possible and can grow wild in our homes. In order to see this all come to fruition, we try out the things that we've read can help, or the tricks passed down to us from family members or friends. As new plant parents, that guidance is vital. If it wasn't for those tricks passed my way from family and friends when I first started my journey in greenery, I probably wouldn't have had the success I had coming out of the gate with plant care. These tricks per se, some might even call "hacks." They are simple, yet creative ways to help you keep your plants thriving but also to style them in a way you might not have thought of before. This chapter will provide you with the hacks every plant parent should have in their back pocket. Here, we'll discuss hacks such as how to combine different types of plants in one pot, how to style the top of a pot, and how to create proper drainage. These tricks will help you along the road to becoming better at caring for your plants, but will also help to make indoor gardening more exciting and fun. It's the latter that I live for the most. So for my friends here in the green-loving community, from me to you, here are some wild hacks!

WATERING YOUR PLANTS WHILE AWAY

If you turn to the *Wild Rants* chapter (p.168–195), you'll see where I talk about the need for a plant sitter when going away on vacation, and if you're planning to be gone for more than a few days, you might want to head there first. But here, I want to give you a few cool hacks for how to water your plants when you're taking that weekend getaway and have plants like ferns or calatheas that require a bit more attention. These simple tricks can not only help you while you're away, but also if you just want a little freedom during the week from watering daily. Here are two tricks you can use to water your plants while you're away.

WATER SPIKES

For this hack you'll need a water spike and an empty glass bottle. Using a water spike to help keep your plant friend hydrated is pretty simple. Water spikes are devices that you stick into the soil which slowly release water. You can find them made from glass, ceramic, and even terracotta. I like to use the terracotta spikes just because they sit low and color-wise fit nicely among my terracotta planters. Once you've inserted the spike in the soil, take your glass bottle, fill it with lukewarm water, and slowly tilt it into the spike. Flipping the top of the bottle over into the spike in this way will stop the water from running out. With the terracotta spike, because terracotta is porous, the water will slowly make its way through the spike and to the soil and roots of your plants. Now, while you're away, that *Alocasia* 'Portodora' can stay evenly moist without you having to worry. Water spikes are great because they are inexpensive and can be found in most garden stores and online. This also gives you a way to repurpose used bottles. I try to consider the color and shape of the glass bottle so that I'm not taking away from the styling of my space.

THE ROPE TRICK

For this hack you'll need a vessel filled with water, a bit of yarn or rope, scissors, a small weight or stone, and a chopstick. First, fill any vessel you like with water. The longer you plan to be away, the more water you'll need. Then, cut yourself a bit of yarn or rope. Make sure to cut enough to reach from the bottom of the vessel to the center of the pot, because one end will go inside the vessel and the other end will be buried in the soil of your plant. Next, tightly tie and knot the stone or weight to one end of the rope. Drop the stone into the vessel, then take the other end of the rope and tie a knot. Place that knot on top of the soil and with the chopstick, press the knot and rope down into the center of the pot. Believe it or not, your role is complete. Now it's time for physics to get involved. Capillary action will now take place, as the water will slowly creep its away along the rope and into the soil. You can also create this process by sticking the rope through the drainage hole of your planter. I use this method sometimes when watering plants that I don't want to have wet topsoil.

CUTTINGS AS BOUQUETS

Few things can brighten up a space visually and aromatically like a bouquet of freshly cut florals. A bouquet of flowers is like the cherry on top of a sundae or the sprinkles on top of a cake. They are a last added touch of goodness that brings a living space together. Recently, floral design has become extremely popular, with many designers using their creativity to construct what I can only call floral sculptures—works of art that bloom from their vases and use their base, whether that's a dining table, console table, or windowsill, as a pedestal to show off their beauty. I've never been a huge fan of florals, but over the past five years I've come to love the King Protea (*Protea cynaroides*). I guess the reason I rarely bring cut florals home is because they are here today and gone tomorrow. The lifespan of a cut flower is short, so that moment of enjoyment flickers out like the embers of burning wood floating into the night sky. I lean more toward lush plants to make my spaces feel warm and inviting because I know with proper care I can keep them alive year after year, meaning that vibrancy inside remains constant. There's also those moments, come spring and summer, when you're gifted with a bloom from your houseplants, a reward for providing them with water and the right type of light. That being said, I started thinking about creative ways to take cuttings from plants and styling them in vases, so they could not only continue to live on for a good amount of time while in water, but also eventually develop roots to then be placed in a new planter in your home. I'm talking about propagation.

Here you could use your creativity and plant knowledge to take cuttings from plants and arrange them in vessels to design a beautiful bouquet that can sit atop your dining table. But in order to do so, there are a few things you'll need to know. When it comes to propagating your plants, it's all about knowing where to make that initial cut. After that, it's about understanding the best light to place them in. And finally, letting the vessel do a lot of the talking. So, with that said, here are a few tips on how to go about creating your own cuttings for bouquets.

UNDERSTANDING WHERE TO MAKE THE CUT

STEM CUT This method of propagation will be used on many of your vine-like, trailing plants. Think about any of your plant friends in the *Philodendron* or *Hoya* genus. If it has a vine, it will be propagated using the stem-cut method. The identifier for where to make this cut will be the node. The node is that little bump on the stem of the plant. For your philodendrons, it'll be right below every leaf, where it meets the stem. On your hoyas, you can find the nodes all along the stem. You'll then want to take a sharp, clean pair of shears and cut below the node or nodes. You can now place this cutting in water, making sure that the node is always submerged. It's from this node that the roots will develop. If you come across a vine-like plant that doesn't have a noticeable node, like a creeping fig (*Ficus pumila*) or string of hearts (*Ceropegia woodii*), just cut anywhere on the vine, remove the lower leaves, and submerge that vine in water. Over time, nodes will emerge and from those nodes roots will develop.

LEAF CUT This method of propagation isn't used as much as the other two because there aren't many plants that can be cut and propagated from a single leaf. The plants you'll have the best success with using this method are *Peperomia* plants, snake plants (*Sansevieria trifasciata*), ZZ plants (*Zamioculcas zamiifolia*) and various succulents. This method is simple, as the goal for best results is always to make the cut on the plant where it's green. Because these plants are typically green all over, you can make the cut wherever you like. You'll want to take a sharp, clean pair of shears and make a 45-degree-angle cut, again for surface area and root growth. Then take the cutting and place it in the vessel, making sure that the cut end is always submerged in water.

TIP CUT This method of propagation will be used on many of your tree-like plants. From your fiddle-leaf fig (*Ficus lyrata*) to the rubber plant (*Ficus elastica*), and the draecenas to the aralias, anything with a tree-like, woody base can be propagated using this method. The important thing here is identifying where on the branch to make the cut for the best results. Starting at the base of the branch where it's hard and tree-like, make your way up until you find the spot where it starts to turn a bit brownish green. From that spot, all the way up to new growth, is where you'll want to make your cut. Your success rate here is about 90–95%. You'll want to take a sharp, clean pair of shears and make a 45-degree angle cut. This angled cut allows for a larger surface area to be exposed to moisture, giving more room for roots to grow. Once you've made that cut, place the cutting in water. Like the node of stem-cut plants, with tip-cut plants, you'll want to make sure the cut end of the branch is always submerged in water.

SELECTING YOUR VESSELS

CLEAR GLASS A clear glass vessel is best for propagating because it will allow more light to come in, helping your cutting to grow healthier roots, faster. It'll also allow you to see when the water starts to get gross. To ensure the health of your cutting and its roots, make sure to replace the water in your vessel once it starts to become murky.

AMBER GLASS Amber glass is also a good vessel for propagating because it will allow light in, although it is less effective than clear glass because you cannot tell whether the water has turned a bit brown or murky inside. For the sake of your bouquet, I'd suggest replacing the water once a week.

SOLID VESSELS A solid vessel, like a ceramic or porcelain vase, will bring a lot of sexiness to your bouquet, but it can make the growth of your roots drag on a bit longer.

LIGHT

FINDING THE RIGHT LIGHT To ensure your cuttings not only stay alive but continue to grow healthy roots, the best spot for them will be a place that gets bright indirect light. This is where you'll see the fastest and healthiest root growth. Medium/indirect light is good as well, but the root growth will take a bit longer. If you try to force cuttings into low-light areas of your home, your success rate will drop. If the cuttings do survive, the root growth will take even longer to develop. The one thing you'll definitely want to steer clear of is direct sunlight. Direct sunlight will heat up the water in the vessels, killing the roots, and can also burn your plants' foliage.

Now with this knowledge, go forth and create a bouquet to set the mood of your space, while at the same time growing new plant babies to add to your collection further down the road.

PREPARING YOUR PLANTS FOR A MOVE

As your plant collection grows and the plants settle into the spaces they are currently in, it becomes increasingly difficult to figure out what you would do if you ever had to move house. That's currently a concern of mine, given I have an upcoming move and will have to pack up 200+ plants. One of those plants being my good friend Frank, my 14ft (4.25m) fiddle-leaf fig tree, that I've had since 2014. When I got Frank I was living in New Orleans, Louisiana. There I moved him from one apartment to another and then, in the spring of 2015, we packed up and moved to Baltimore, Maryland, where I live today. I had become a dedicated plant parent only a year before this, so I wasn't well versed in how to properly prepare myself for this move. Not only was I leaving with Frank, but with 60 or so other plants as well. This was pretty much trial by fire, but every plant made it safely to the new home. During this move I took note of how I could have better prepared myself, so that when I made another move, there would be less stress and more efficiency. Here are my tips on how to prepare your plants for a move.

THINKING AHEAD

DRY SOIL One thing you'll want to make sure to do is to let your plants dry out before the move. This means that if you plan to move a week from now, it's best to let the soil start to dry out so that the plant and pot are much lighter and easier to transport. Trying to move planters that have wet soil in them can be grueling and makes your move much more difficult. If you're dealing with a plant that needs to be watered more often, try putting a water spike in the soil and placing a bottle filled with water on top. That will help keep your soil moist without it becoming too wet and heavy.

NURSERY POTS While you might be excited to get that new plant into its new pot, leaving your new friend in its nursery pot could be beneficial, especially if you tend to move around a lot. Plants do perfectly fine in their plastic nursery pots and you won't have to repot them until their roots start growing out of the drainage holes. A plant in a nursery pot is much lighter than a plant in a heavy ceramic pot. Believe me, I learned the hard way. Trying to move Frank in his ceramic pot created such a mess and extended the moving day. Now, if I'm bringing plants into a workspace that I know I might not be renting for a long time, I'll place my plants that are in their nursery pots inside a larger pot. That way, when it's time to move, I can just pull them out and move them with ease.

PACKING YOUR PLANTS

COVER THE POT In many cases you'll first need to make sure you protect the pot. Using bubble wrap to protect it from getting damaged is the first thing you should do. Next take a plastic bag or sheet and tightly wrap that around the top of the pot, taping it down to make sure none of the soil spills out. Just make sure you don't cover the drainage holes. You want to make sure the roots can still breathe.

BRING THE FOLIAGE TOGETHER A lot of your larger plants have been enjoying their best lives and spreading their arms out in your home. But now you need to get them out of the front door, into a truck, and into your new place. The best way not to break off one of their limbs is to gently pull them inward. To do this, use gardening Velcro or string to slowly bring the branches or stems together, and tie them up. This will force the branches and stems to go more in the up direction than out.

PROTECT THE FOLIAGE Now that you've tied up your branches or stems, you'll want to protect the foliage from getting torn, punctured, and in some cases, sunburned. Using kraft paper, pull off a piece that's long enough to wrap around your plant. Once you have it around all of the foliage, use tape to secure it. This also helps when you're moving during the colder months of the year. It'll help protect the foliage from frost bite.

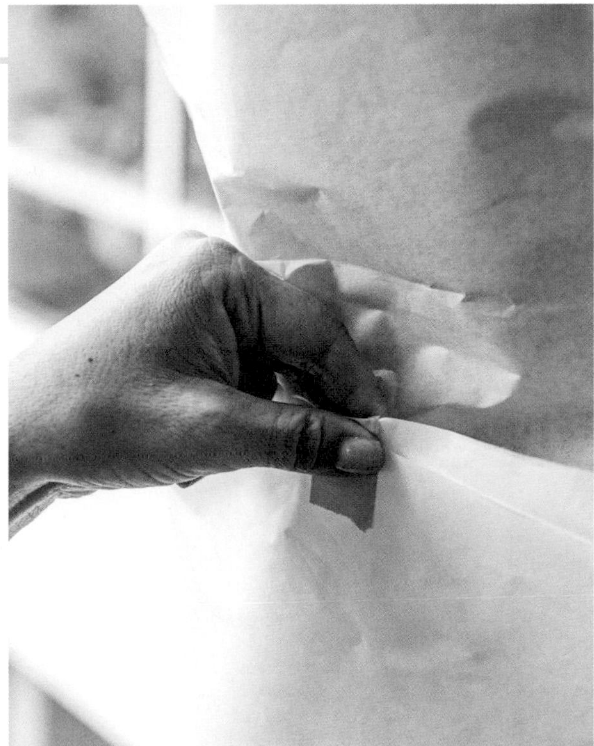

TIMING IS EVERYTHING

BEST TIME TO MOVE Just like everything else involving plant care, the best time to move your houseplants is during spring and also fall. When the weather is anywhere between 65–85°F (18–30°C), it's less harmful exposing your plants to the outdoors. While we can't all predict when we'll have to make the transition from one space to another, doing it during this time of the year would be best. If your plants are exposed to low or high temperatures for too long it can severely damage the foliage and kill the entire plant after a while.

INSIDE A VEHICLE When moving, you want to consider how long your plants will be inside your moving vehicle. If moving during the summer or winter, remember to treat your plant like all the other living things in your life. They will die in a vehicle baking in the sun or left out in the cold. So if moving them long distance, when you stop to go inside a place, whether that's a restaurant or hotel, bring them with you or roll down windows and open doors so they can get not only air but also a little light. Honestly, the safest way to travel long distance with a lot of plants is to hire a plant-moving service. And if you think you won't be able to keep all of your plant friends alive during the journey, there is no shame in giving them to a friend or family member instead.

These tips should help you get your plants from point A to point B safely. Hopefully you're moving into a home with more light and a lot of southern exposure. I'm wishing that for you all.

HANGING PLANTS WITHOUT CREATING HOLES

We all want to fill our homes with as many plants as we can possibly care for. That's the sign of a true plant parent. We find creative ways to display our plants and one of those ways is by hanging them throughout our home. But to do so could mean creating many holes in the walls and ceilings throughout your space. As a self-proclaimed plant hoarder, this can create a bit of a holey situation when you want to have plants hanging everywhere. Living in apartments or spaces I've rented for all my life has forced me to become more creative in how I can avoid making holes in my walls, when it comes to hanging plants. So here are some quick hacks you can use when hanging plants, so that you can get your security deposit back.

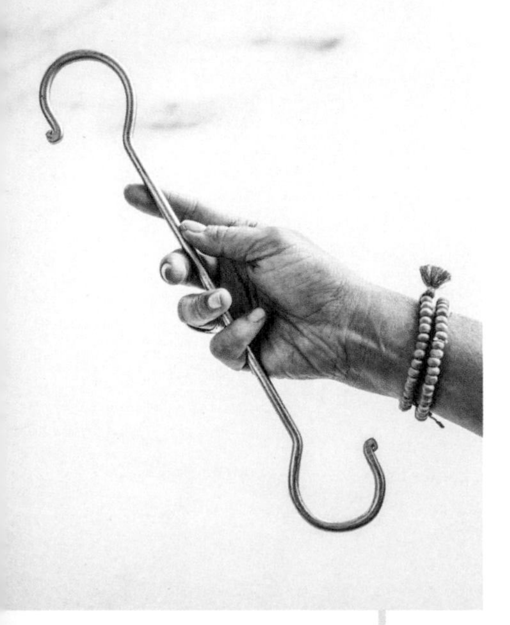

CURTAIN RODS

These are already drilled into your walls, so why not hang your plants from them? Look, I understand you might want to close your curtains from time to time, but why not have a "living" curtain? For me, this is the perfect way to cover my windows to deter unwanted gazes from outsiders while also providing me with hanging greenery throughout my home. This doesn't require me to make any new holes in my walls and places the plants in really great light. The one thing you have to make sure of is that the rod is drilled into the wall using screws and anchors and that it can hold the weight of the plants you're hanging from it. To make it a bit more stylish, I'll hang cool hooks from the rod, and and then hang plants over the other end of the hooks. You can find these hooks online or in local boutique plant stores.

TENSION RODS

I've been using tension rods to help in my plant care routine since my very first plant. Tension rods are great because they aren't permanent and can hold a good amount of weight. I use a tension rod over my kitchen sink to hang my plants on after I water them and I also utilize them in my skylights. Placing tension rods in your skylights not only gives you a way to hang your plants without creating holes in your walls, but now that you have plants hanging in your skylight, they can also help diffuse or dapple the light that streams into your space when the sun is at its highest point.

CHAIN OR ROPE

When possible, I'll avoid drilling holes in my walls and ceilings by using the metal pipes or beams that are already well secured in the space to hang chain or rope around. The main thing here is to make sure to ONLY use the pipes and beams in your space that are solidly attached to the structure of the building, not pipes that are functioning as your sprinkler system, plumbing, etc. Once you've identified a beam that can work for you, the world of hanging plants is your oyster. The one great thing about using chain is that you can easily secure it and adjust the height by using screw-locking carabiner rings. Using chain not only allows you to adjust the height of the plant easily, but also allows you to hang multiple plants from one chain.

The idea here is to limit the amount of damage being done to your space while building your indoor oasis. Believe me, I've created a few holes here and there, but when I can avoid them, I try my best to find a creative away to do so.

MAGNETIC HOOKS

These are game changers, especially for those living in lofts that have exposed metal beams. You can purchase magnetic hooks that can hold anywhere from 25–100lb (10–45kg). Yes, you read that correctly. All you need to do is attach the hook to a metal surface in your space and the magnet will do the rest. This allows you to hang a variety of plants throughout your home without drilling holes and makes it easy for you to remove plants when it's time to water them. It's a clean look that works in any space with metal surfaces.

BUG OFF

The one thing that you can almost guarantee when bringing plants into your home is that at some point you will see a bug or two. That's just a part of "bringing the outdoors in." It's like wearing white and being surprised that you got a stain on it. One begets the other. When it comes to plants and bugs, it's kind of like that. Some bugs are destructive to the life of your plants, while others are just an annoyance. That's probably why the number one question I get from most plant lovers is how to deal with bugs. My hope is to try my best to get on top of the issue before it becomes a real problem. So here are some tips to limit the number of critters you see crawling or flying around your plants.

SHOPPING

You could have had plants in your home for years and then one day, out of the blue, you have an infestation. Well, that plant you just brought home from the plant store could just have been the culprit. When you're shopping for a new plant friend to bring home, you want to give it a good look over. I talk about treating your plants like pets in a later chapter. Look at it as if you've just taken your dog for a long walk through tall grass. Just as you would check your pup for ticks, you're going to want to get in there and check those plants for bugs. Just because you're getting the plant from your favorite plant store doesn't mean that it isn't already infested with bugs. So before you purchase it, make sure to check on top of the leaves, below the leaves, and around the top of the soil for anything that looks like it's not a part of the plant. While some bugs move around a lot while on your plant, others are motionless and can easily be missed. If you're uncertain whether what you're seeing is a bug or not, ask an assistant at the nursery for help. The first step in limiting bugs in your home is making sure you're not the one bringing them in.

THE TOP FOUR BUGS TO LOOK OUT FOR:

1 MEALY BUG — White and cotton-like

2 SCALE INSECTS — These look lifeless with shells that look like droplets of tree sap

3 SPIDER MITES — Small and create webs around the edges of your foliage

4 GNATS — Small and fly around wet soil

CHECKING IN

While every plant parent should expect to see a bug here and there, you won't run into an infestation overnight. It's not like there's one bug one day, the next day two, and then the third day a thousand bugs. That's not how it works. But if you're not paying attention to your plants, that's when it will happen. The best way to keep your plants as bug-free as possible is to check in on them at least once a week. And by checking in I mean, again, giving them a good look over, as you did when shopping for them. If you want to make sure your plants are thriving and that you don't find yourself with a plant that is covered in mealy bugs or spider mites, this work is necessary. And while you're in there, go the extra mile and wipe the leaves down. I try to do this every other week to keep me on the lookout for bugs, but also to remove any dust that has built up on the foliage. Removing that layer of dust can help bring more light to the foliage, making them vibrant, and can help give your plant back its natural shine. Sometimes I'll place a droplet of mild dish soap in a gallon (4.5 liters) of lukewarm water and I'll use that with a soft cloth to wipe the leaves down. While removing dust, the mild dish soap can also deter and kill bugs. After, I'll come back and wipe the leaves down again with a clean cloth.

CEDAR WOOD

Nature has a way of protecting itself. They say that when you cut the leaves of a philodendron, the scent they release is the plant warning others around it that danger is among them. I love that. So when it comes to cedar, that's what makes them a great bug repellant. They release a scent that bugs like termites, moths, mosquitoes, and gnats don't like. When trying to rid your plants of these bugs, an easy method is to place cedar chips, sawdust, or even pencil shavings on top of the soil. Yes, I said pencil shavings. Those No.2 (HB) pencils we all use to write with can be used to deter bugs from coming around your plants. I like using pencil shavings because one, I'm repurposing the shavings I already have from sharpening my pencils, and two, it looks cool (see *Styling the Top of the Pot*, p.138).

BASE WATER

Watering your plant from the base will allow the soil and roots to pull moisture in without getting the leaves and the topsoil wet. When it comes to gnats, they love being around damp soil, so if your topsoil isn't as wet because you're now watering from the base, you'll see fewer bugs flying around your plant. When watering from the base, it's important to only allow the water to sit in the base tray or the pot to sit in water for 30 minutes. After 30 minutes you'll want to dispose of any remaining water or remove your plant from water.

OLD TRICKS

Like many of us, I've been handed down tips and tricks to deal with things that can occur in our lives. One trick that I've been shown for dealing with a few pesky gnats, is to place a small bottle cap full of apple cider vinegar or leftover coffee on the topsoil or near your plant. This will draw the gnats there, for them to die, while the soil of the plant dries out.

CREATING LIGHT FILTERS

When it comes to sunlight streaming through your windows, while all of your plants need light, some of them won't enjoy direct sunlight. Especially if it's afternoon direct sun. Being able to create filters to protect the fragility of your plants' foliage from direct sun is a must. This is the reason why between the glass ceiling of a greenhouse and the plants below there will always be a thin sheet of frosted glass or gel. These filters help to diffuse the direct sun, cut down its intensity, and create an even glow of bright white indirect light.

So, if your home is flooded with afternoon direct sun and your plants are sitting in it, you might want to create a filter to protect them. Remember, direct afternoon sun can burn the foliage and ultimately kill your plant.

Here are a few hacks for you to create a filter at home.

FROSTED GLASS OR GEL

If you've ever been in a bathroom with a window, I'm sure that window was either completely frosted or had a portion of it that was. Most people will frost the glass in a bathroom for privacy while still allowing daylight to come in. The gel that you use for this can also be purchased and placed on other windows in your home. If you have a dedicated area of your home where most of your plants reside, you could gel those windows and give that room a greenhouse feel instantly. Using frosted glass or gels is more of an investment, but the clean look makes it well worth it.

BUBBLE WRAP

Yes, just like you'd use bubble wrap to protect your fragile items from getting damaged in the post, you can also use it to protect your plants. Now don't go wrapping the foliage. Place the bubble wrap evenly over the window, taping the sides or creating a frame and placing it inside. While it might seem cheap, you can finesse it in a way that adds texture and style, and ultimately keep your plants thriving. The wrap will diffuse the sunlight coming in, creating the glow your plants will love.

SHEER CURTAINS OR BLINDS

These are more common and fit in any home. What's nice about using sheer curtains or blinds is that you can easily control how much light comes into a space. If you have a western-facing window, you could have the curtains open during the morning hours and then close them in the afternoon.

CREATING DRAINAGE

If you've been following me via social media, seen any of my plant care videos, heard me on a podcast, or read my previous books, you've probably heard me express over and over again how important it is to have a drainage hole in your planter. So I apologize because I'm going to say it again here: having a drainage hole in your planter is key to the health of your plant. I need to drill this into your heads, pun intended (by the way, if you're reading a book about plant care, regardless of the writer, know that all puns are always intended). A drainage hole allows the water you're giving your plant to have an escape route from the bottom of the pot so that the roots don't remain submerged in water, which would eventually cause the roots to rot and over time kill your plant. And you're not a plant killer, right? You're a plant lover! And as plant lovers we must make sure our plant friends, kids, whatever you're calling them, are properly cared for. To do so, when potting your plants in new planters, it would be germane to choose one with a drainage hole. One great thing about plants becoming so popular right now is the vast variety of beautiful and stylish planters being produced. But I often wonder why so many of them come without a drainage hole. Oh believe me, I'm just as frustrated as you are. With all of that said, if your planter doesn't come with a drainage hole, well, you're going to need to drill one yourself. Now I know what you're thinking. You have a plant in a pot without a drainage hole and what you did was create a little buffer zone in the bottom of the pot with stones and charcoal, so that when you water your plant, water runs through the soil and roots and settles into this buffer, keeping the roots from sitting in water. Well if you've read *Wild at Home*, you know I'll agree that that's an OK alternative. And it's just "OK" because what you said will happen. That water will make its way through the soil and roots and finally settle into the buffer zone. But there will be a time when that water doesn't evaporate as fast as it did the previous week, so when you come to water again, this time the water will go down into the pot, filling up the buffer zone and making its way up to where the roots are. And this isn't what we want. We don't want to play guessing games with the health of our plants' roots. We want to know for sure that the water going into the pot has a way to escape and creating a drainage hole is the best way to do this.

So here are a few tips on how to go about that.

WITH AND WITHOUT A TRAY
Here, my smaller rubber plant (*Ficus elastica*) and burro's tail (*Sedum morganianum*) sit in the studio window, catching a little fresh air. Since the burro's tail doesn't have a base tray, it's taken to the sink for watering.

CREATING A DRAINAGE HOLE

1 Get yourself a drill bit that works for the material your pot is made of and for the size of hole you'd like to make.
2 Safety always comes first, so make sure you are wearing protective eyewear.
3 Turn your planter over and make an X with masking tape to mark the spot where you're planning to drill the hole. The tape not only provides you with a guide to where to make the hole, but also helps to keep the pot from breaking.
4 Place the drill bit in the center of the X and start drilling. Be careful not to put too much of your weight behind it. If it doesn't seem to be going through with a little force, check to see if you have the correct bit.
5 Once the hole is created, wipe off the dust, remove the tape, and prep your plant for repotting!

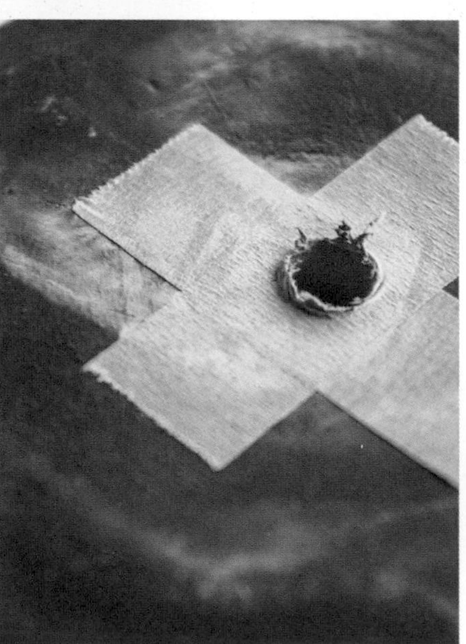

Now I know some of you are going to say, "Well Hilton, my planter was too expensive for me to try to drill a hole in it and potentially cause it to crack or be destroyed completely." I completely understand those fears. So if you are going to place a plant in a pot without a drainage hole, I hope that plant is still in its nursery pot, sitting inside of your more stylish pot, and when it's time to water it, you're taking it out of that stylish pot and carrying the plant to the sink, tub, or wherever you water your plants, watering it there, letting it dry, and then placing it back in your pot without a drainage hole. Because like I've said over and over again, you don't want plant roots sitting in water.

BASE TRAYS

Here's a thing you might not have thought of: once that drainage hole is created, you're going to need a place for that runoff water to collect in. And hopefully that's not your wooden floors or rugs. You're going to need a base tray. Now because it's mostly the more stylish planters that come without a drainage hole, if you do drill holes to create drainage, you're going to want to place a base tray beneath it that doesn't strip away any of the sexy from the planter and plant. Another hack I'd suggest here is using stylish plates, quiche dishes, or trays as your base tray. Honestly, it can be anything that has a raised lip around its edge high enough to retain any of that runoff water. As someone who enjoys repurposing and upcycling, I know those old plates and trays in your cabinets are great for this. I'll often plan trips to thrift stores and flea markets just for these sorts of things. OK, so remember, once that water enters that base tray, let it sit there for about 15–30 minutes, giving the soil and roots time to pull in any water they hadn't had the chance to grab onto as it made its way through the pot. After 30 minutes, if there's still water in the tray, take your plant off the tray, dispose of that leftover water, place the tray back in its spot, and finally place the plant back on top of the tray.

STYLING THE TOP OF THE POT

With the plant being the star of the show and the planter being the outfit (the dress if you will), that your plant is wearing, sometimes you want to add a little bling to pull it all together. I mean, come on, if you're going to name your plant, you might as well make sure it looks good. In this case, covering the top of the soil can make your plant stand out and but it can also help with the life of your plant.

It's all about being creative while also considering the type of plant you're working with and the individual the plant is for. You could use Lego bricks or marbles for kids, or for someone more spiritual, you could consider using crystals.

STONES

Using stones to style the top of a pot is a classic look. I'm sure you've seen this done, but possibly haven't given it a shot yourself. If not, you should! Especially if you have desert plants like cacti, succulents, or snake plants. One reason why is that it just looks good. For the most part, these desert plants grow vertically, exposing the soil, so why not just enhance their look by dressing the top of the soil with something more appealing. Another reason why it works with desert plants is because they require less water than other houseplants, so while I've mentioned that you'll want to check the moisture level of the soil before watering, with a desert plant, given that you won't be watering it often, you'll never have to worry about getting your finger through those stones in order to check the soil. Overall, it's a really great look that makes your desert plants stand out and brings in that extra touch of the outdoors.

MOSS

Not only does moss help to make your terrariums look amazing, utilizing reindeer moss, sphagnum moss, and sheet moss to dress the top of your pot can do the same for your potted plants. In the case of mosses, I'd recommended only doing this for plants that like their soil to be evenly moist between waterings. The reason I say this is because the moss will help retain the moisture that you've given your plant and this isn't something you'd want for a plant that likes its soil to dry out. Adding moss is a simple and fast way to bring a little more green into your home, but it doesn't mean that you have to make it look simple. Because preserved moss comes in so many different colors and looks, you can get a little creative in the way you style it.

USED TERRACOTTA POTS

If you have cats, you might have memories of when your fur friends got a little too excited and knocked over one of your plants. That has definitely been the case in our household. There have also been times when a plant's roots will grow through the drainage hole, so that the only way to repot the plant without damaging the roots would be to break the pot. In a situation like that, instead of tossing that terracotta pot in the trash, I'll wash out the broken pot to clean it up and then take a hammer to break it into smaller pieces. It's these pieces that I'll use to decorate the top of my planters. Terracotta is a porous material, so it'll pull moisture away from the soil. So for that reason, I only use it for the containers of plants that like their soil to become dry between watering. This is a really cool way to upcycle your broken pots while also providing your plant with a fresh look!

Lastly, another great reason why you should dress the top of the pot is if you have a pet that likes digging into soil or using it as a toilet—not only will you have a beautifully styled plant, you won't have soil all over your floors or a dying plant in your home.

COMBINING PLANTS IN ONE POT

When it comes to bringing plants into a space, one way that I let my individuality and creativity shine is in how I style different types of plants in one pot. Yes, that is a thing! I mean you could place your Swiss cheese plant (*Monstera deliciosa*) in the same pot as your 'Marble Queen' pothos (*Epipremnum pinnatum*), *Philodendron* 'Xanadu', and your *Philodendron* 'Pink Princess'! And do you know what happens when you take all of those plants and place them in the same pot? OK, well I'll tell you. The answer is simple... you make your friends jealous! And isn't that what it's really about? Your friends come over to your home and they see this amazing grouping of greenery in one single planter and they lose their minds! And with that you feel great! But in order to have this kind of fun, you must first know what types of plants can thrive in the light your space has. Then, it's all about knowing what types of plants require their soil to have the same sort of moisture when it comes to watering them. As I've mentioned before, when it comes to watering a plant, that's going to be based on the moisture level of the soil. Then it's understanding the type of light they need. So, when combining plants in the same pot, you want to work with plants that are like-minded. For example, you could place cacti and succulents in the same pot with a snake plant (*Sansevieria trifasciata*) and a ZZ plant (*Zamioculcas zamiifolia*), or pot a lemon button fern (*Nephrolepis cordifolia* 'Duffii') in the same planter as a medallion calathea (*C. roseopicta*). Once you have a good understanding of your plants, that's when you can pretty much go wild!

For my workspace I knew I wanted to put together a grouping of plants that would not only work together color-wise but would also grow nicely out of my pedestal planter. Over time, I could imagine some foliage growing upward and some crawling down the side of the pot. So to make this all work, I went with two types of philodendron, a 'Rojo Congo' and a 'Lemon Lime'. Given that philodendron are vine-like plants, I could see them cascading down and alongside the pot as they grew. I also added a *Dracaena* 'Lemon Lime' to work the vertical space. With all three plants requiring their soil to be half dry between waterings, placing them in the same pot is possible.

MY HOME OFFICE
Having plants in a home office is important as it helps you to be more creative. In my office, I've mixed things up by growing different but similar-minded plants in the same pot.

POTTING THE PLANTS

Just like any plant, you want to make sure you pot in a new planter that is 2in (5cm) in diameter larger than the previous pot or nursery pot. Since I'm potting three plants in one pot, I placed them all inside the new pot just to make sure that there would be enough room. Once I saw that they could all perfectly fit inside my planter, I added a soil medium that works for the types of plants I have. With philodendrons and dracaenas, using a soil that is well-aerated is helpful, so I used a mix that's 80% potting soil and 20% vermiculite or perlite. I then took the plants out of their nursery pots, gently loosened the soil and roots, and then placed them with intention inside the planter. Understanding that the 'Rojo Congo' will grow wide, I placed it at the back of the pot, making room for the 'Lemon Lime' philodendron to still get the exposure it needs. I then added my topsoil, patted it down (not too tight), cleaned off any dirty leaves, placed it where I wanted it to live and, lastly, watered it. Now I have a grouping of plants in one pot, ready to show off! Oh, and don't worry, your friends will love you for it.

SOME OF MY FAVORITE PLANT COMBINATIONS

Swiss cheese plant (*Monstera deliciosa*) + *Philodendron davidsonii* + tree philodendron (*Philodendron bipinnatifidum*)

String of pearls (*Senecio rowleyanus*) + string of bananas (*Senecio radicans*) + burro's tail (*Sedum morganianum*)

Burgundy rubber plant (*Ficus elastica* 'Burgundy') + 'Rojo Congo' *Philodendron* + *Philodendron* 'Pink Princess'

Maidenhair fern (*Adiantum raddianum*) + lemon button fern (*Nephrolepis cordifolia* 'Duffii') + creeping fig (*Ficus pumila*)

Watermelon peperomia (*Peperomia argyreia*) + *Peperomia albovittata* 'Piccolo Banda' + baby rubber plant (*Peperomia obtusifolia*)

Stromanthe sanguinea 'Triostar' + medallion calathea (*Calathea roseopicta*) + *Calathea 'Vittata'*

REPURPOSED

Whether you call it upcycling, reusing, or repurposing, having the awareness to find treasure in another person's "trash" is important for the health of our planet. Don't get me wrong, I like the look and feel of something shiny and new just as much as the next person, but having the ability to look at a used item and find a cool and creative away to give it a second life is everything. My personal style has always been a little vintage mixed in with a little of the new. It's the balance of the two that I find so appealing. While styling my space, I've also tried to think of ways to repurpose items I had that weren't getting much use, or things that would normally be tossed out with the trash. A couple of examples: the propagation wall I have in my home all came about when I had the idea to use a spice rack as a propagation vessel; and while shopping at a flea market, I purchased a vintage wooden tiger head sculpture and used it as an air plant (*Tillandsia*) holder. Sometimes you just have to look a bit deeper to see the potential an item might have in your home. In repurposing, you not only keep that item out of the landfills, you also make your home more unique and stand out. Here are some ways you can repurpose items in your home or things you might find in thrift stores, to make your home more of an indoor jungle.

DISPLAYING CUTTINGS
Placing cuttings from plants in repurposed vessels is an easy way to add a splash of color to your home. Here, Jasmen is taking a cutting from the rubber plant, *Ficus elastica* 'Burgundy', to place in a champagne glass.

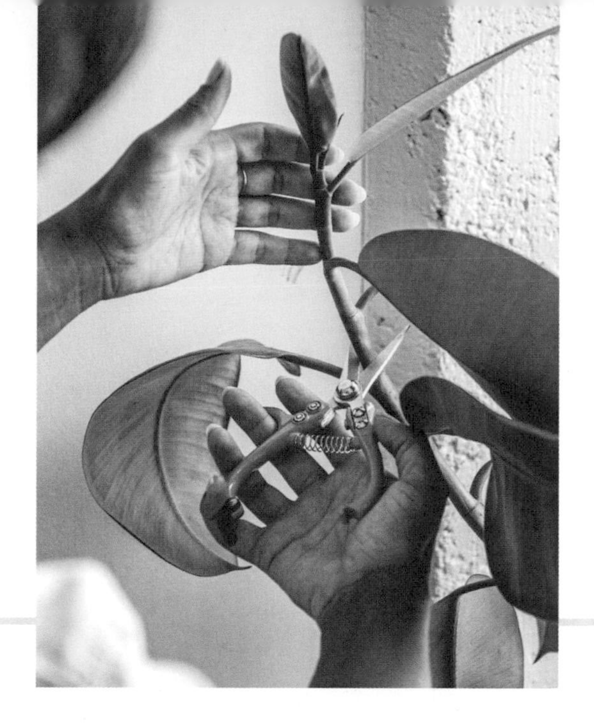

NURSERY POTS

Every new plant you bring home more than likely comes in a plastic nursery pot. At the farms where your plants are coming from, they place them in these pots as they grow and then they'll make their way to your local plant store or nursery. While most decide to instantly repot their plants in new pots and toss the plastic nursery pots in the trash, I prefer to keep them around. One, they are perfect for when your propagated cuttings develop roots and you want to pot them, and two, they are also useful when you want to leave a plant growing in nursery pots. Instead of buying a new, larger pot, once my plant has outgrown its nursery pot, sometimes I'll just repot it in a larger nursery pot. This is very helpful for those that have larger plants and want to be able to move their plants around the home more easily. Plants in nursery pots are much lighter and easier to move than those potted in ceramic or terracotta planters (see *Preparing your Plants for a Move*, p.120). Lastly, if you'd prefer not to keep your plants in the nursery pots they come in, once you remove your plants, try to take the pots back to the plant store or nursery you got your plant from. I'm sure they can find a way to repurpose them.

DR J. HOSTETTER'S
STOMACH BITTERS

VINTAGE GLASSWARE

For me, every glass object I come across has the potential to become a propagation vessel. I mean, why not? All you really need is a clean glass vessel filled with water and a cutting from your favorite plant. So when I stumble across vintage bottles or clear glass vases, I see how they can house the cuttings of my plants. What's great about glassware is that there have been so many different shapes used to make these objects over time. From pieces of glassware used in the field of science to those created for beverage consumption, they all have a beauty and uniqueness of their own. Propagation leans heavily on the side of biology, so placing a cutting in a beaker just feels right. What makes using vintage beverage containers so great is that the labels were either painted on or molded into the glass. So now your vessel has a little extra pop of color or nostalgia connected to it. But maybe that vintage/used look isn't your thing and you like to keep it classy or modern. Well in that case, I'm sure you have nice glassware in your cabinets that rarely gets used. For example, we have more champagne glasses than any normal household should have, so to mix a little new with the old, I'll take a champagne glass and use it as a propagation vessel. Now you've classed it up and truly have something to celebrate! With all of the different types of glassware out there, you really could create your own unique propagation station that is to die for.

CERAMIC VASES OR MUGS

Just like clear glassware, there are so many items made from ceramic that can be repurposed and used as planters. Whether it's a bowl, cup, mug, or ceramic vase, in my view, they all have the potential to hold a plant. While this look is more clean and classy than using tin cans, it's more difficult to create the proper drainage necessary to give a plant the best life possible. To do so, you'll want to create a drainage hole at the bottom of the vase or mug (see *Creating Drainage*, p.134), or if that's not possible, create a little buffer zone at the base of the ceramic piece with stones and horticultural charcoal. Here I've used a vintage mug, shaped as a lady in lingerie, to pot my cactus in. While the mug already has a lot of attitude, placing a cactus at the top makes it more of a statement piece. To me that statement is: "You can look, but you can't touch." Sounds about right.

TIN CANS

Unlike today, where most of our packaged goods come in cardboard or plastic containers, back in the day most of it came in tin cans. Everything from coffee to crackers would come in tin cans with the labels painted on them. Since we moved away from that, there are many vintage and thrift stores that have these cans taking up room on shelves. Well, why not give them a home and put them to use in your indoor jungle? These tin cans could make the perfect planter. In doing so, you'll want to create a drainage hole at the bottom of the can (see *Creating Drainage*, p.134) or if that's not possible, create a little buffer zone at the base of the can with stones and horticultural charcoal. What's nice about using these tin cans is that they stand out against the other planters you'll have in your collection and, through the labeling on the can, you can show your sense of style. Here I've used a vintage beer can from a brewery that once operated out of my city. While it might not mean anything to you, for me its sparks a bit of hometown pride that makes it mine.

CREATING YOUR OWN POTTING MIX

Have you ever thought about how important the potting mix you're providing your plant is? If not, then let me tell you... it's very important. There are two reasons for repotting a plant. One is to help fertilize your plant and the other is to help it grow, when the plant's roots have outgrown its pot. When you bring your plant home it already comes packed in soil, with the nutrients and fertilizer it needs for at least a few seasons. This is why when you first purchase a plant, it's not necessary to fertilize it until the following spring. Over that time period, through watering, the soil will gradually lose the nutrients it has and these will need to be replenished. While some people like to add fertilizer to the soil that's there, others prefer just to add fresh soil to their plant. In this fresh soil, you are giving the roots back the nutrients, or let's say food, your plant needs. So to fertilize your plant is one reason why you might be adding new soil. But the most common reason for providing a plant with new soil, the one that most of us find ourselves doing annually, is when it's time to repot. And like I mentioned, you will know the plant is ready to be repotted when its roots start to creep their way out of the drainage hole of the pot. When it's time to repot, you can't just grab any random bag of soil from the plant store or use what's left over from your shed. This is why it's important to really know the type of plant you have, because different plants require different types of potting mix.

Providing your plant with the wrong potting mix will lead you and your plant down a dark and stressful path. Let's say, for instance, you have a cactus. You're going to want to place your cactus in a potting mix that is fast-drying and well-aerated. We all know that a cactus is a desert plant, which means it likes dry soil. When repotting cacti, you want to make sure you have additives in the soil that can help remove as much moisture as possible, and quickly. You might find your typical "cactus and succulent" potting mix to be sufficient, but you may be required to add a little something extra to help dry the soil a bit more. It's the same thought process in reverse for a plant like a fern which likes its soil to remain moist. You would never give a fern and a cactus the same potting mix. One would become dry and die because of under-watering, while the other would die because of overwatering. Remember, it is the moisture level of the soil that will tell you when to water your plant, so providing your plant with the right potting mix is everything.

To help you give your plants the soil they need, here are my tips for creating the perfect potting mix for some plant families.

IMPROVE THE POTTING MIX
Adding bark to potting mix can both aerate the soil and help your plant's roots to breathe. For plants that need their soil to stay moist, mixing in some bark is the perfect addition to the potting mix.

PERLITE
SOIL
PEAT MOSS
VERMICULITE

ADDITIVES YOU MIGHT CONSIDER ADDING TO YOUR POTTING MIX

PERLITE These are the white little clumps you might find in a bag of organic soil. I like to purchase full bags of this and just have it at home, ready for whenever I might need it. Perlite helps to pull moisture away from the soil. It's also helpful in improving drainage. If you have a potting mix that is becoming too compacted and doesn't drain out properly when watered, definitely add perlite.

VERMICULITE This potting mineral is more lightweight and light brown in color, so when mixed into soil it's not going to pop out as much as perlite. It's a great addition when looking to help aerate the soil, but it also helps to retain the moisture in the soil. It also improves drainage. Add vermiculite if you have a potting mix that is becoming too compacted and doesn't drain out properly when watered.

PEAT MOSS This is a dead material which is great for plants that require their soil to stay moist. The moss helps to retain moisture in the soil and gradually releases it between waterings.

BARK MULCH This is mainly shredded pieces of bark, which helps aerate the soil and improve drainage. Bark, like peat moss, helps to retain moisture for your plants as well.

POTTING MIX FOR SPECIFIC PLANTS

PHILODENDRON AND FICUS

These are plants like your fiddle-leaf fig (*Ficus lyrata*), *Ficus benghalensis* 'Audrey', rubber plant (*Ficus elastica*), golden pothos (*Epipremnum aureum*), *Philodendron* 'Rojo Congo', and *Philodendron davidsonii*.

When adding new soil to these plants I like to make sure I always have a bag of organic potting mix ready to go. Since philodendrons and ficus mostly like their soil to be moist for a day or two and then become dry at the top half of their pots, I make sure my mix is 80% potting mix and 20% perlite. This is all based on whether I've placed these plants in porous containers. If I've placed my ficus in a ceramic glazed pot, I might add 10% more perlite to help remove the moisture in the soil.

FERNS AND CALATHEAS

These are plants like your Australian tree fern (*Dicksonia antarctica*), maidenhair fern (*Adiantum raddianum*), *Calathea orbifolia* or calathea medallion (*C. roseopicta*).

When adding new soil to plants like these, which need their soil to stay evenly moist, I make sure my mix is 70% potting mix, 20% peat moss, and 10% vermiculite or bark. While I'd never place any of these plants in a porous container, if you decide to do so, please remember that you'll need to water your plant more frequently.

CACTI OR SUCCULENTS

These are plants like your apple cactus (*Cereus repandus*), crown of thorns cactus (*Euphorbia milii*), jade plant (*Crassula ovata*), or snake plant (*Sansevieria trifasciata*).

When adding new soil to plants that want their soil to be dry for long periods of time between waterings, I make sure to always have a bag of cacti and succulent potting mix readily available. In most cases, adding more perlite to these mixes is helpful. Cacti and succulents grow best in porous containers but if you decide to place them in a plastic or glazed container, you'll want to add more perlite to the potting mix.

VIBE SETTERS

If you have read my previous books or seen any home that I've styled, you'll know that I'm a big believer in creating a mood or vibe for a space so that those who enter that space feel instantly at peace or better yet, at home. The "vibe" permeates throughout a space and allows those in it to feel connected in some way, whether that's through the music you're playing, the scent in the air, the glow of a candle's burning wick, or the styling of the greenery there. Creating that vibe is everything, as connects us all. At the end of the day, the idea is to make a space feel more warm, energized, and vibrant. You can change the energy in a home just by burning a bit of sage, playing some music (my recommendation: "Low End Theory" by A Tribe Called Quest), or placing a few candles about the room. Well at least that's how it works for me.

You see, it's not going to be the same for each individual because while burning a stick of *palo santo* can look the same in any home, the scent it releases might trigger a unique memory in each individual. For my wife and I, burning *copal* in our home takes us back to Tulum, Mexico, where we got married. During the ceremony, the shaman burned *copal* to cleanse the air but also to keep mosquitoes away. It's the memory of that *copal* smoke slowly filling the jungle air, floating right above blades of Chinese fan palms (*Livistona chinensis*), and creating a canvas to give structure to the sunlight's rays, that takes us back to Tulum each time we burn it in our home. Now when others come over, they get to feel a bit of that journey with us, or have their own memories triggered. That's what setting a vibe is all about.

STRINGS OF LIGHTS

These aren't just to decorate your Christmas tree or make your patio bright. Bringing strings of lights into your living space among your greenery is a true vibe setter. The warmth of the bulbs adds a nice little touch against the green of your foliage and at the same time, strings of lights just scream out party!

SAGE

Burning sage has always felt like such a peaceful, Zen-like ritual to perform in a home. They say it not only helps to cleanse the air but also to deter bugs. To fellow plant lovers, I shouldn't have to say more. But I will. For those that lean a bit more toward the spiritual side, they say it helps to remove negative energy and we can all use something that can help with that from time to time.

PALO SANTO

Burning a bit of *palo santo* is great because, like your plants, it is also supposed to help clean the air. I find myself burning it when we have guests and for some reason, on rainy days. During my plant-care sessions it's become a part of my ritual to burn a *palo santo* stick, put on some music, and tend to my plants.

COPAL

Like sage, *copal* is used to deter bugs but is also known to cleanse the air and bring good energy into the home. It's one of my favorite types of incense to burn at home and I couldn't recommend it enough.

CANDLES

Scented or unscented, lighting candles is an instant vibe setter. They're the mascot for the word "mood." While they can illuminate any dark space, candles are the main attraction at any birthday celebration, the waving flame of a lost loved one, and the warmth of romance.

SEASHELLS

When traveling I love bringing back a little treasure from my journey that can always transport me back there when I see it in my home. Collecting seashells, if permitted, while you're on a beach getaway has always been special to me. They instantly remind you of those moments your toes were deep in sand or washed over by the ocean's tide. While they can be great to style in a home just by themselves, I love using them as votive holders, *palo santo* holders, and sometimes holders for my air plants.

THE PLANTER STANDS ALONE

When it comes to plant styling, it's not about simply pushing plants into a space. It's about making a space come alive, stylishly, through the lens of your creativity. So not only are you picking out the perfect plant for a space, but also the perfect planters. And the selections can feel like reaching the bottom of Netflix. Believe me, I've tried that before. While you've heard me mention over and over again how important it is that your planter should have drainage, it's just as important to have a place for all of that runoff water to exit into. In many cases this place would be a base tray. But not every planter comes with a base tray and sometimes the planter just looks too beautiful to place just any random tray beneath it. Therein lies the dilemma. Do you just place your plant in a pot without a drainage hole or do you water it just enough so that no water comes out of the drainage to ruin your floors? Whenever I'm using a planter that demands that it stands alone, without a base tray beneath it, I'll use these two hacks to make that possible, while still making sure my plant gets the proper drainage it requires. FYI, these hacks can only be achieved by leaving the plant in its nursery pot.

HIDE THE BASE TRAY

This is just as simple as it sounds. Instead of potting the plant in the planter and placing a base tray below it, you leave the plant in its nursery pot, place a base tray inside the planter, and then place your plant inside. Now your plant can sit inside your beautiful planter while appearing as if it doesn't have a base tray. This look is not only cleaner, but also more stylish and modern. When it comes to watering your plant, do just as you would with any other plant in a pot with drainage, water until it comes out into the base tray but don't let that water sit there for more than 30 minutes. Remember, if there's still water in the base tray after 30 minutes, you'll want to dispose of it. Meaning you'll have to take your plant out of the planter and dump the run-off water out of the tray. If the plant is too heavy to lift, take a towel and place it around the base of the pot to absorb the water, or use a baster to suck it out.

CREATE A BUFFER ZONE

If you use a planter without a drainage hole or that has a stopper in its drainage, you can create a buffer zone using stones or bricks in the bottom. With your plant still in its nursery pot, place it on top of the stones or bricks to keep it off the bottom of the planter. This buffer zone will now give the runoff water a place to go when you water your plant. Just remember not to allow too much water to fill up at the base of the planter. You don't want your plant sitting in a pool of water. This hack will also allow you to place smaller plants in larger pots. Over time, as the plant grows, it can grow into these pots nicely.

What's great about this hack is not only does it allow your planter to show off without a base tray, it'll also make it so much easier for you to move your plants around. Keeping your plants in their nursery pots and not potting them in a planter gives you the flexibility to switch them out from pot to pot. It'll also make it easier if or when you have to move out of your current home or space. Plants are much lighter when moved in their nursery pots than they are when potted in a terracotta or concrete planter.

LEVEL UP

Can we all agree that in having plants, we have become just better people in general? I think I can say that about myself. I can say in slang terms, I've "leveled up" when it comes to my own personal growth. When it comes to plant styling, you want to level up as well. Not just in your care process for your plants, but in the actual way you position them in your home. In *Wild at Home* I wrote about creating levels with your plants, but now I want to go a bit deeper. While I wrote then about using unexpected surfaces, like your dining table or kitchen island, to bring plants higher into a space, there are other ways to get your plants to stand out from the rest by using some of the smaller surfaces you have around the house. For me, I try to think about how to best give each plant or object its time in the light. I mean that literally and metaphorically. The goal is to, in a way, create small stackings of platforms, so that even if two plants are exactly the same size, you can place one on a higher level to separate it from the other, while making it more visible. I envision this being the most useful approach on a windowsill or in a plant corner. Here are some ways you can level up.

MARBLE OR STONE SLABS

There have been times when I've come across marble cutting boards or slabs at thrift stores and purchased them just to create levels for my plants. Or a time when a friend would be doing renovations on their home and have a few stones left over from a project, so I'd take one to use for my plants. What's great about using marble or stone slabs is that they add texture and color, which separates them from the surface you are placing them on. Unlike wood, which can be porous and allow moisture to move through it, if you place a wet planter on top of stone or marble, the water won't make its way through, so it won't damage the surface below. There are so many amazing surfaces to choose from. I've even seen some people stack their plants on top of cinder blocks.

vaporisateur
after first spray

OLD BOOKS

I know this is a weird thing to write in a book of my own but, yes, you can and should use old books to create levels for your plants. I mean, who are we kidding, I've seen many of you stacking your candles on top of copies of *Wild Interiors*. Oh sorry, that was actually me who did that. Whatever, you know what I mean. The idea here is to reuse those books once you've pulled the knowledge from them and have them lying around, to become surfaces for your plants. There are plenty of used bookstores and thrift stores where you could go and grab as many old books as you like to create your look. And my books are the only books that you have my personal permission (once you've absorbed all the knowledge within) to use as plant stands.

WOODEN CUTTING BOARDS

If you're like me and love hosting parties, I'm sure you've got your fair share of wooden cutting boards. They come in so many cool shapes and sizes, and make the perfect serving surfaces for your cheese and charcuterie. But there seems to come a moment when wooden cutting boards have had their day and it becomes time to throw them out. Instead of doing that, why not repurpose them as small plant stands? This is where I'll get creative and stack them up in my windowsill and place my plants on top. Now you've created less waste and have simultaneously given your home a fresh new look. Sometimes I won't even wait until my cutting boards are too gross to cut or serve on and will just purchase cool cutting boards with the sole intention of using them as plant stands.

WILD RANTS

When it comes to styling a space with plants, for me, it's not all about the amount of plants you bring into the space that makes the difference. I'm not one that you'll hear say things like "The more the merrier" or "You can never have too many plants." What truly makes the difference is the care you put into the plants you do have. A single plant can make a cold, stale space feel warm and inviting. So why does it feel like everyone is racing to fill their homes with as many plants as possible? I do feel a bit responsible for this. I have on multiple occasions throughout my books mentioned all of the plants that I have, but do know that that's because I feel capable of caring for the ones I have. There's been more emphasis lately on the number of plants people have in their homes rather than the quality of care going into those plants. Believe me, it's not a race. No one is going to come knocking on your door once you have 500 plants to present you with a gold thumb award, I can promise you that. Let's be clear, plants aren't props. They're not a new pair of shoes or a vintage desk lamp. Regardless of how trendy plants might be at the moment, they are living, breathing things that should be treated and respected as such, so understanding how to care for them properly is important. Your plants don't have the ability to walk themselves over to a sink for more to drink or make their way to a brighter window in your home to get better light. They need you to be aware of their needs and provide them with such necessities. The goodness that you put into them, they'll return back.

I'm sure most of you have heard about the studies that NASA has done regarding bringing plants indoors. They talk about

how plants can clean the air and add more oxygen to a space. Those are all wonderful things to think about but I'm sure it would take an abundant amount of greenery in a small space for an individual to feel any sort of tangible benefit here. We've surrounded ourselves with over 200 plants in our 1,000 square-foot (92 square-meter). apartment and I've never once come home and thought, "Oh wow! All of this oxygen! I'm so glad I have so many plants!" Not once has that happened. I do feel a sense of calm and serenity while among my plants, but that has very little to do with how many plants I have. I also have a large pit bull named Charlie who walks throughout our home and when doing so, he moves through the space like a cloud of stink. The plants we have aren't cleaning the air around him. We burn candles and *palo santo* to clean the air in our home. I can't truly say clean air and added oxygen are tangible benefits of having plants. I can't feel that. What I can feel, what is tangible, is the energy that comes through me when I see a plant thriving. Watching as a plant you've spent so much time caring for unfurls beautiful new growth will make you feel something. Something raw and weighted. For me, it's empowering because I know I helped make that happen. I knew what light was best for the plant and placed it there. I knew what soil medium that plant needed and surrounded its roots with it. And I thought about the material the planter I was placing it in was made of, so that it could better benefit from that. When you see a plant thriving you walk around more confident and proud. It simply makes you feel good.

I've heard people say, "Plants make people happy" but that's only if you're putting in the work to see them happy. Plants can also make people sad. If you're like me, when I see a plant not doing so well, I feel awful. Knowing how to provide your plants with the care they need can be easy if you just put in the work. Hear me out, there is no such thing as having a brown thumb for particular types of plants or having a brown thumb in general. I'm not judging you, but those are all excuses. People who consider themselves as having "green thumbs" are just individuals who decided to put in the work. They made an effort to give the plant what it needed to thrive and saw their care pay off. What's happening is that many plant lovers are making decisions on what plants they purchase on the popularity of a plant or how rare it is first, and then thinking about where it can fit in in their space. They're not thinking about what's best for that plant but selfishly about their own wants. I found myself falling victim to this earlier in my journey in greenery. Then there was a point when I realized that I needed to make a change. I needed to let the light coming into my home guide my decisions on what plants to bring in. Look, I get it. I know how hard it is to go into a plant store or nursery and not leave without every single plant they have. Oh, I know the feeling. I compare it to going to an animal shelter. Of course you want to take all of the animals there home, but you don't. For one, no one wants to be known as the crazy cat person, and two, you take the time to consider what you can care for, if you have time for it, space for it, and then, only after thinking about all of that, plan accordingly. And as I like to say, you need to plant accordingly. Once you're able to see plants as living things, while you may love an Australian tree fern (*Dicksonia antarctica*), with your schedule and your ability to be a bit forgetful, you decide to leave it at the plant store for someone who has time to water that fern every single day. I mean, honestly, ferns are just the plant versions of puppies.

THE POWER OF LIGHT

LET IN THE LIGHT
On the right, the afternoon gaze of the sun shines bright in the palm house at Kew Gardens, London. When it comes to greenhouses, it's all about letting in the right amount of light.

Light is everything to every plant. Honestly, it's the most important thing that a plant needs. Well, that and a nice hug here and there. While I'm sure some of you are thinking water is the most important thing, you have to understand that there are some plants you'll bring into your home which will go dormant during the colder months of the year, meaning you'll need to water them much less frequently than you have been. You see this with the trees outdoors or the perennials in your garden. Frangipani (*Plumeria*), cacti, and many other plants that you have indoors will also go dormant. So while plants need water, some won't need as much during certain times of the year. But when it comes to light, they'll need the same type of light year around.

The light you give a plant will change everything about it. The more light a plant gets, the more vibrant its foliage will be, showing off that plant's true color. You can easily see this with the golden pothos (*Epipremnum aureum*), which gets its name from the goldish tones that appear in its foliage. The more light this plant gets, the more gold you'll see. When placed in lower light, the gold isn't as prominent. Light also helps flowering plants bloom more often. If you own a crown of thorns cactus (*Euphorbia milii*) and want to see it bloom, give it more light. It's light that changes the shape of your plants as well. The more light a plant gets, the larger its foliage will be and the less light it gets, the smaller its foliage will be. Take the *Monstera deliciosa*. The more light you give this plant, the larger the foliage will grow and the more holes and splits its leaves will produce. Yes, those splits and holes... they are due to light. When exposed to bright light, the leaves of the *monstera* grow larger and to ensure that its leaf friends below get enough light, the leaves will split apart to allow light to shine through it. There's something so poetic and beautiful about that. And speaking of the *monstera*, if they get a lot of great light, they will bloom for you. They are a flowering plant. But don't hold your breath. You'd probably have to live in a greenhouse to see that happen.

Light is so important that I hope from this moment forth, before you leave home to purchase a plant, you'll find the spot where you're looking to style a plant and then use the light that spot gets as your compass to guide and dictate what types of

plants you can have there. As plant people let's make sure we are only bringing in plants that can thrive in our homes, not just survive. That's very important. Like all living things, plants are just looking for a place to show off their true potential. I know in my past, as a plant hoarder, I'd find myself at a plant store, and I'd grab whatever plant I thought looked cool and bring it home. Not knowing much about plants other than they needed light and water, I'd find a spot with "light," only to come back days later to see the plant freaking out. This would then cause me to freak out. I'd move the plant to a new spot, the plant would freak out again, and so on and so forth. It's a vicious cycle and during this merry-go-round, that plant would suffer until I finally figured it out. I've learned from this and hope you can as well. Let's put an end to that and know exactly where our plant friends can live in our homes based on the light we can provide for them. And when I'm talking about understanding light, I'm talking about knowing what direction your windows are facing because whether you have north-, east-, south-, or west-facing windows, this will give you a great understanding of what plants to bring into your home.

The light guidance I'm about to give you relates to homes in the northern hemisphere, so if you live in the southern hemisphere these directions will need to be reversed. While northern-facing windows give you an exposure that's anywhere between indirect, medium, or low light, an eastern-facing window should give you a morning direct sun, that is a cool direct sun. Not cool meaning "dope" but cool meaning temperature. That morning direct sun is actually OK for your houseplants. But let's not get that AM direct sun confused with that PM/afternoon direct sun you get from western-facing windows. Afternoon direct sun is more intense. The kiss of afternoon sun on most of your indoor plant friends can be brutal. It can burn their foliage, which shows itself in orangish/brownish spots on the leaves that are facing the window, and over time can kill your plant. As plant lovers we want to protect our plant friends, so if you have plants that can't tolerate that afternoon exposure, you're going to need to create some sort of filter (see *Creating Light Filters*, p.132). That filter against the direct sun will create a glow that is called bright indirect light. That's what all indoor plants will thrive in and your south-facing

BASKING IN THE SUN

Above, our kangaroo foot fern (*Microsorum diversifolium*) hangs out, basking in the late evening sun. While ferns aren't fans of direct sun, they do well when kissed by morning direct sun. Just make sure never to place them in a spot that gets too much afternoon direct sun.

windows will give you plenty of that. If you're unfamiliar with the term bright indirect light, then let's get familiar. A plant that is sitting in bright indirect light is a plant that has the ability to take in the filtered light of the sky without being exposed to direct sun. Imagine being in a field on a cloudy day. The sun is above those clouds, the clouds now operate as our filter, absorbing all the intense heat from the sun, balling it up nicely, like a perfectly packed snowball, and reflecting on to you a nice even glow of bright indirect light. If you're trying to figure out if your plant is sitting in bright indirect light, whether they're in a windowsill, hanging in a window, or sitting on top of your dining table, if that plant is fully exposed to the sky but not to the sun, it's sitting in bright indirect light. So an eastern-facing window could possibly get direct AM sun but then get bright indirect light in the afternoon, while a western-facing window can get bright indirect light in the morning and then direct sun in the afternoon. It all depends on the direction your windows are facing and where your plants are in relationship to that window.

The size and what's outside of your window matters as well. Just because someone has a south-facing window doesn't mean they have bright indirect light flooding in. That window could have a large tree outside it, blocking much of that great filtered light, or it could be facing a large building. Consider the surroundings outside your windows and take all of this information with you when shopping for plants. Providing an assistant at a plant store with this information can help them guide you in the direction of plants that can live in the light exposure your home has. Just as a quick note, if they talk to you about "low-light-loving" plants, please know that there is no such thing as a low-light-loving plant. Those rattlesnake calatheas (*C. lancifolia*), ferns, snake plants (*Sansevieria trifasciata*), ZZ plants (*Zamioculcas zamiifollia*), Pothos, Chinese evergreens (*Anlaonemas*), etc., don't love low light, they only tolerate it. They'd thrive in bright indirect light just like all your other indoor plants. But I get it. I'm a plant lover as well. I want to spread that greenery throughout my home and with some corners not being as bright as others, a ZZ plant might have to take one for the team.

RIGHT PLANT, RIGHT PLACE
The dog's tail cactus (*Selenicereus testudo*) hangs in our western-facing window, enjoying some afternoon sun. Cacti and succulents are the perfect choice for western-facing windows given that they thrive in direct sun.

CAT PEOPLE VS DOG PEOPLE

WATCHING AND WAITING
Above, our cat Isabella sits patiently as she ponders whether she should find a nice cozy spot in the bedroom. Right, our dog Charlie gazes out of the window hoping he'll be able to meet the squirrels that play across the road.

Have you ever heard me tell my story about the difference between cat people and dog people? No? Well, gather the kids around the fire because it's story time with Hilton Carter! Yeah, I know this is a book about your green babies, not your fur ones, but bear with me for a moment and I promise I'll make this all connect. Most plant people are pet people anyway, so I believe this will hit all the right places. I for one am a dog person. I have a dog named Charlie that I adopted six years ago. My wife, well, she's a cat person. When we met she had two cats, Isabella and Zoe. We all now live in the same house as one big happy family, but that's not the story. The story is breaking down the difference between the person that decides to own a cat and the person that decides to own a dog. While both are animal lovers, they tend to make these decisions for specific reasons.

So as I mentioned, my wife is a cat person and loves everything about those little fuzz balls. From the way they chase reflected light, to their screams when they want a little attention. I've caught her once or twice meowing back. She feels connected and bonded to them and they feel the same about her. The question is, is she a cat person because she likes cats or because of the type of person she is? Not to make a blanket statement about those of you that have cats, and I know most of the plant community does, but there are some things I've noticed about my wife that she benefits from because of having cats. As a cat person, you get a little more freedom to live your life as you always have, while still getting the love from a warm animal. For example, my wife loves to sleep in whenever she possibly can. She could sleep all day if you let her. Having cats, she can do so because there's no rush for her to get out of bed to care for them. More than likely they're sleeping themselves. The house is their entire world. Cats don't need to go outdoors to relieve themselves. The cat person places a little litter box somewhere in the home for them to go, and once a day or every other day, whenever is good for them, they go and clean it up. So the cat person gets to sleep in as late as they want. As a dog person, that's not your life. A dog person knows what they're signing up for when they decide to get a dog. You can kiss all of your weekends of sleeping

in goodbye. As a dog person you know you'll have to wake up every morning, in most cases early, and take your pup for a walk, and you'll be doing this three to five times a day. And that's regardless of how cold or hot it is outside, no matter if it's raining, snowing, or if the Earth is on fire, you'll have to take them for a walk.

Cats are also different types of eaters than dogs. Cats seem to understand that if they don't eat all of their food at once, it'll still be there when they come back to it, so the cat person gets a little leeway if they ever get a bit forgetful and forget to feed their cats. My wife likes to go away for weekend getaways and before we take off she'll leave out a large bowl of food, a few bowls of water (this is because our cats enjoying knocking over bowls of water), and a fresh litter box, knowing that the cats will be fine taking care of themselves for a couple of days. For a dog person, this could never be the case. While the cats are fine to take care of themselves, I have to find a dog sitter to take Charlie for the weekend so that he can get the care he needs. A dog person is required to go the extra step, even if that means paying to have their pup cared for while they're away. Dog people find themselves forced into being more involved with the wellbeing of their animal. And when it comes to plants, there are certain similarities and considerations one must think about before choosing a plant. Because some plants are like having a dog, while others are like having a cat.

The idea is to be self-aware enough to know where you fit in here. Are you someone that's a bit forgetful or just likes the idea of having plants but doesn't want to give up their weekends tending to them? If so, for that person, they should lean more

toward "low maintenance" plants like ZZ plants (*Zamioculcas zamiifolia*), cacti, succulents, snake plants (*Sansevieria trifasciata*), dumb canes (*Dieffenbachia* 'Camille') etc., that won't require much of their attention but will make a space feel lush. These plants are more forgiving and if you're someone that travels often, you won't need to get a plant sitter while you're away. You'd want to stay away from ferns, calatheas, anthuriums, alocasias, and certain palms.

Or maybe you're a "dog" person. Someone that is up for tending to plants on a daily basis and being focused on their care. This sort of person can welcome all plants. While I know it's difficult not to bring every plant home from the plant store, you have to be self-aware enough to hold off. Just think about them as cats and dogs. Of course, we'd all want to bring every single cat and dog home that we saw in an animal shelter, but we don't. We consider what we have the ability to care for, our home environment, our work schedules, etc. and make a decision based on that. Do the same when it comes to bringing home plants and you and your new plant friend will be better off. So I guess the question is, what are you... a cat person or a dog person?

PETS AND PLANTS DON'T ALWAYS MIX

Indoor jungles and indoor pets basically go hand in hand. Just as you should know what types of plants to bring into your home based on the type of light you have, you need to consider what types of plants you can bring in based on which ones are toxic and non-toxic to pets. While I've lucked out and have pets that don't find plants interesting at all, not everyone is as lucky.

THE PLANTER

As a plant stylist, I see the planter, AKA the pot, as the dress or the suit we place our plants in to give them their personalities, but also to show off your own sense of style. When it comes to the planter, you have to get it right. And when I'm talking about getting it right I don't mean, you know, picking out a planter just because it matches the color of your throw pillows or the design of your wallpaper. While I do think that's cool (I did mention I'm a stylist), that's not the first thing I think you should be considering when placing a plant in a pot. What you should first consider before placing a plant inside of a planter is the material that your planter is made out of and how that can help the life of your plant. Because I do want you to understand that all planters are not created equal. Let's say you were going to place a rattlesnake plant (*Calathea lancifolia*) in this beautiful planter you just purchased from a store. You made the decision to get that particular planter because of its design, color, and texture. And who can blame you?! Those are three of the main decisions I make when purchasing a new planter. You got your new planter home and you thought the vibrant colors of the rattlesnake would look great in it. But here's the thing, you didn't consider the fact that your planter is made out of clay. And clay, terracotta, concrete, and most woods, are porous, breathable. Meaning they'll pull moisture away from the soil of your plant, causing the soil to dry out faster. And that rattlesnake plant wants its soil to stay evenly moist, never dry. The same goes for ferns, anthuriums, alocasias—anything that wants its soil to stay evenly moist will not appreciate being in a porous container. You should place plants like these in a glazed ceramic pot or plastic container. These containers lack the ability to let air move through the pot, which will help keep the soil of your plant moist.

Save all of the porous pots for those plants in your collection that want their soil to become dry between waterings. Your cacti, succulents, ficus, and most philodendrons. Just make sure you don't place them in plastic or glazed containers. Well, that's unless you take the appropriate steps to get your soil right (see *Creating Your Own Potting Mix*, p.152). I know some of you are thinking, "Why can't I put my cactus in a plastic pot? Isn't the nursery pot it came in plastic?" That's a valid point. Yes, those nursery pots your plants come in are plastic but here's the thing, the farmers that grow those plants

WORKS OF ART
A mix of planters can bring originality to your plant collection and a stylish look to your home. Here, I have displayed a few from ceramicists whose work I really love.

in those pots make sure to give the plants the right soil medium and additives that will allow the soil to dry up faster while being in their plastic nursery pots. For instance, if you ever look at the soil of a cactus inside a nursery pot, it generally remains dry because they make sure to use a fast-drying, well-aerated soil. So, if you were going to decide to repot your cactus in a glazed ceramic pot or plastic container, just make sure that you're using additives like perlite or vermiculite to help dry your soil out faster. It's these considerations that will help set you and your plant up for success. Once you have an understanding of how the material your planter is made out of will react with the soil inside said planter, then you can have fun with it and start thinking about whether the planter works well with the color of your couch and throw pillows.

CLAY OR TERRACOTTA PLANTERS

These pots allow air to move through them. While this will cause your soil to dry out faster, it also allows the roots to breathe. The clay pulls the moisture from the soil and holds it in the clay, drying out over time. As this process repeats over and over again, watering after watering, you'll start to notice some residue appearing on the outside of the pot. This is normal. It's just the sediment from the water, fertilizer, and soil.

Plants you should style in pots like these: Cacti, succulents, ZZ plant (*Zamioculcas zamiifolia*), snake plant (*Sansevieria trifasciata*), and ponytail palm (*Beaucarnea recurvata*).

CONCRETE AND STONE PLANTERS

Just like clay, these pots are porous and allow air to move through them, causing the soil inside to dry out. These planters are great because they are very durable and can last forever. The only negative side to having a concrete or stone planter is its weight. Because of this you'll mainly see them used for outdoor greenery. I like to use smaller concrete planters indoors because I love their raw texture and color against the green of the foliage.

Plants you should style in pots like these: Fiddle-leaf fig (*Ficus lyrata*), rubber plant (*Ficus elastica*), *Ficus benghalensis* 'Audrey', tree philodendron (*Philodendron bipinnatifidum*), and *Yucca*.

CERAMIC GLAZED AND PLASTIC PLANTERS

These pots are perfect for plants that want their soil to stay moist, because they retain moisture. For planters like this it's important to have a drainage hole because you can easily cause a plant's roots to rot if there isn't one. Glazed and plastic planters are typically the ones that will be the most colorful, making them more eye-catching. Just make sure you're placing plants in them that want their soil to stay moist longer or placing additives in your soil that can help strip that moisture away.

Plants you should style in pots like these: Maidenhair fern (*Adiantum raddianum*), asparagus fern (*Asparagus setaceus*), *Stromanthe sanguinea* 'Triostar', medallion calathea (*C. roseopicta*), *Calathea orbifolia*, and *Alocasia* 'Portodora'.

THE PLANT SITTER

Over the past few years of sharing my passion for plants and the plants in my home, many have asked how my wife and I make sure our plants survive while we're away on vacation or while I'm away on a book tour. Well, the answer is pretty simple. Once you grow your plant family to the level that we have, you need a PLANT SITTER. Yes, you read that correctly. I said plant sitter. Go ahead, laugh now but if you don't heed my words, you'll definitely cry later. Just as you'd need a sitter to watch your kids or your pets, you'll need a sitter to come in and care for your plants while you're away. And just as you do your due diligence when selecting a babysitter, you should put the same energy into finding the right person to care for your plants.

How I go about it is by first selecting someone in my life that not only cares for plants but has a true care for me. These people will undoubtedly go the extra mile to ensure the health of your plants because of their love for you. Or you can make the decision based on how you've seen them care for their own plant family. You don't want someone that is struggling with their own plant gang to come and struggle with yours. Find yourself someone that is committed. This person should think about plants with the same respect as you do. Also understand that just because they know how to care for their plants it doesn't necessarily mean they know how to care for yours. Their fiddle-leaf fig (*Ficus lyrata*) could require a different watering cycle than yours, based on where it is in their home. So when you leave the lives of your plants in their hands, set your plant sitters up for success!

For me, the best way to go about making sure plant sitters have the keys to success, is to create a cheat sheet for them that breaks down exactly what plants need to be watered, what plants don't, and how to go about it all. I'm sure many of the pet owners out there do a little of this now. You make sure to tell your sitter how much food to give your little fur buddy, how often to walk them or play with them, and what things to simply avoid. It's basically the same for your plants but they don't need to walk them. Actually now that I said that, maybe they should walk your plants too. I'm sure they get a bit jealous looking out of your windows at all the outdoor plants getting to sway in the breeze and dance in the dappled light. I think I'll take my money tree for a little walk later today. Now I'm on to something. But I digress.

When creating this cheat sheet, I do my best to make it as clear as possible. I start by drawing out which watering can they should use (if you're a plant person like me you probably have multiple watering cans throughout your home). I then start tagging all of the plants in my space with a colored sticker to specify which watering group they are in. GREEN means it's on a seven-day cycle, BLUE means it's a two- to three-day cycle, and RED means DO NOT TOUCH. The ones marked with red are more than likely my low-maintenance plants like my cacti, snake plants (*Sansevieria trifasciata*), ponytail palms (*Beaucarnea recurvata*), etc. Creating this cheat sheet will give them a guide to how to go about not killing your plants and keep you from losing your mind when you come home.

THE PROPAGATION WALL
Our propagation wall is one of my favorite things about our home. I find myself replenishing the water in the vessels every few days to make sure the roots are always submerged in water.

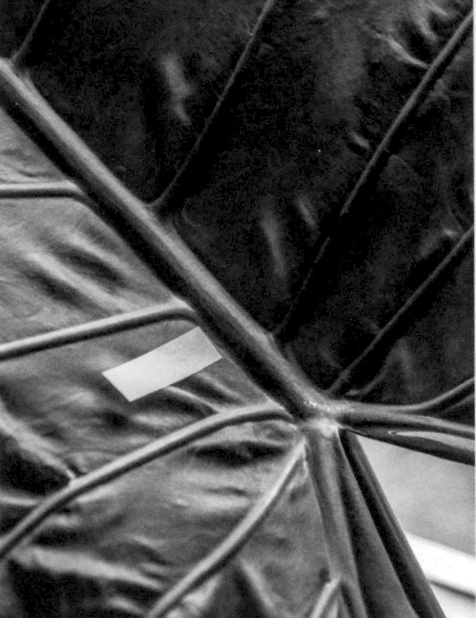

Another good tip to make it easy for them is to group your likeminded plants together. For example, if you have a lot of cacti, instead of just leaving them spread throughout the home while you're away, making your plant sitter have to seek them out, group them all in one area of the home and just tell the sitter this is a DO NOT TOUCH zone.

I make sure to note that all the plants should be watered until runoff water makes its way out of the drainage hole of the pot and into the base tray. It's important that these details are mentioned so that it's clear how much water your plants should receive. I also note if there are any plants that might need to be taken to the sink or tub to be watered and then returned to their spots. I'll break it down based on rooms and the specific plants in that room requiring additional attention. You know, the kids that are a bit more needy. In my case, it's my propagation wall. The water evaporates throughout the week, meaning I always need to make sure the roots or cutting ends of the propagations are constantly submerged in water. So this requires a bit more attention than my potted plants.

Lastly, a day or so before I leave, I invite the plant sitter over to our home and we do a full walk-through so that everything is clear. Not everyone has as nice handwriting as I do (clearly this is a joke). With all of this done, you'll feel more comfortable going away on vacation, not having to worry about your plants and their wellbeing. And like any concerned parent, don't feel ashamed to ask your sitter to share pictures while you're away just in case you start missing your green friends. Have fun and safe travels, wherever you're off to! And don't forget to bring something nice back to your sitter to thank them. I know I won't forget about mine!

ALWAYS CARE FOR YOUR PLANTS

Just like getting a dog sitter, making sure your plant sitter is set up for success can help put your mind at ease when you're away. When I first got into plants, I never considered how important this was. I thought I could just go away and that my plants would be OK. Once I started to make a conscious effort to see plants as living things, I made it my duty to ensure they were properly cared for, just like the other living things in my life.

EXAMPLE CHEAT SHEET FOR THE PLANT SITTER

Use the watering can by the sink.

WATERING LABELS

- **GREEN** Water every 7 days
- **BLUE** Water every 2–3 days
- **ORANGE** Do **not** water

Note: When watering, slowly water plants until runoff water exits out of drainage hole.

MAJOR CALL-OUTS

LIVING ROOM When watering the large fiddle-leaf fig, you may need to place a towel in the base tray to absorb any runoff water.

HALLWAY 1 The propagation wall is BLUE. The tubes will need to be topped off every 3 days.

2 While most of the plants in the window are cacti and won't need to be watered, please take the hanging plants down and water them over the sink. Once they've drained, place them back in the window.

BEDROOM The *monstera* is the only real issue here. The pot doesn't have a base tray, so take the small red bowl I left next to it and place under the drainage hole when watering.

AIR PLANTS Collect them all and place them in the sink filled with lukewarm water. Let them soak for 15 minutes and place them on the drying rack. Once they're dry place them back in their spots.

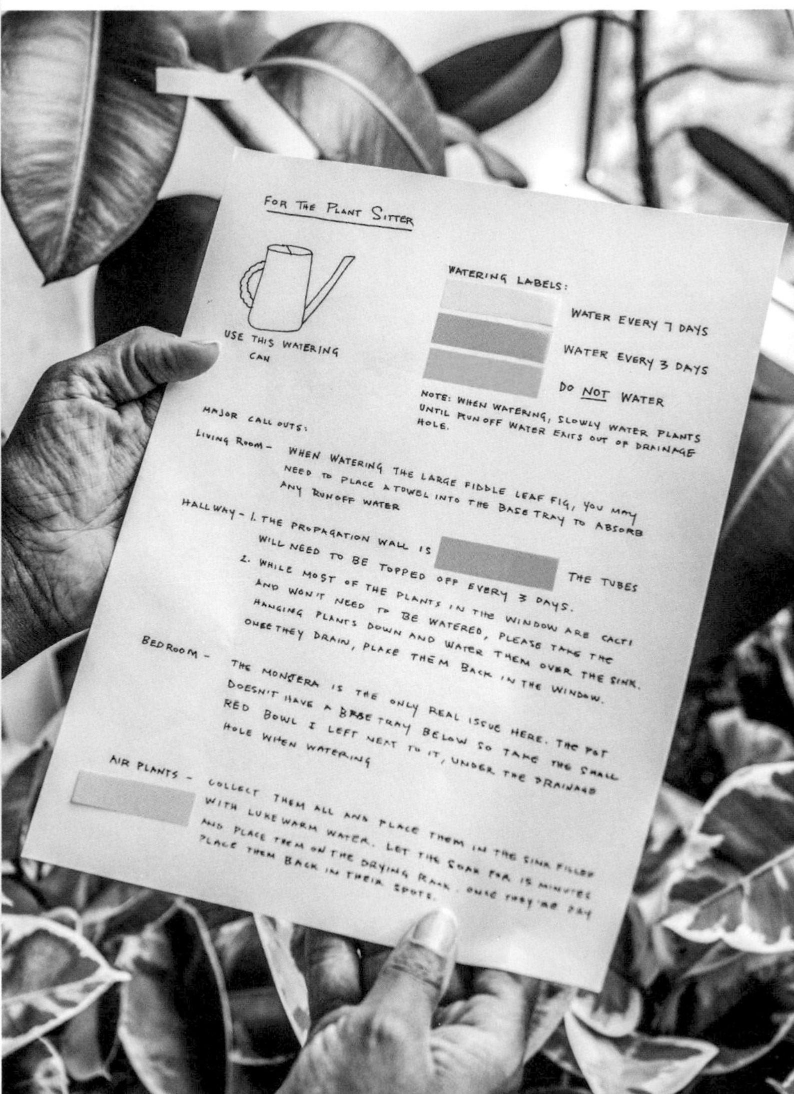

PLANT TALK AKA A LOVE LANGUAGE

Have you ever wondered what the first houseplant was? I know I've spent a good amount of time staring at my plants contemplating this. I mean, it is interesting to think about how this idea of bringing the outdoors in first came about. I figure some caveman, let's call him Hilton (cool name), is sitting in his cave styled with a dinosaur-bone coffee table and palm-tree couch. He's sitting there just staring into space because he has nothing better to do, while eating his leafy salad. He turns and peers out into the field, sees more plants, and thinks to himself, "Hmmm, it might be cool if I brought some of those plants inside. But instead of eating them, I'll just display them near the opening of my cave. Add a little life to this place." Honestly, there had to be a moment in time when this first happened. Well, regardless of when that moment was, not long after, that first houseplant owner started talking to that plant. And let's be real, he probably named it as well. It's this connection with plants that many of us share today. I know I find myself talking to my plants all the time. I use my words to encourage them to grow, to just check in, and to occasionally voice my problems. And I'm not the only one doing the talking, the plants talk back to me as well. While I'm verbally speaking, they speak with their foliage. It's in these moments that you have to be listening.

Plants are talking all the time. Whether that's through color, form, or scent, they are talking to us. When you see a plant with foliage that's turning fully yellow, that's the plant telling you it's receiving too much water, but if that yellowing is just happening on the tip of the foliage and looks a bit orange around the edge, that's the plant telling you the water might have chlorine in it. If the foliage is curling or drooping (fainting), that's the plant telling you it needs a drink. The peace lily (*Spathiphyllum wallisii*) is known for fainting when it's thirsty, making its leaves droop low to the floor. But once you give it a drink, the foliage slowly rises back up. If you see dry brown spots on the edges of the foliage, that's the plant telling you it's not getting enough water. But if you see browning happening in the interior of the foliage, that could be the plant letting you know its roots are rotting and you should check on them to fix the problem. Listening when your plants are

LET YOUR PLANTS TALK
Left, my ming aralia (*Polyscias fruticosa*) lets me know that it has had too much to drink by the yellowing in its foliage. Bottom, my peace lily (*Spathiphyllum wallisii*) makes its leaves faint to let me know it would love a drink right away.

talking to you can make the difference between having a home full of healthy greenery or a lot of suffering plants.

Plants talk to us in how they grow as well. If you pay very close attention to the plants growing in a greenhouse, the one thing you'll notice is that a lot of them are growing straight up, versus to the side, like they do in your home. This is because in the greenhouse, those plants are getting the majority of their light from above, and in your home, it's more directional, from the side. Plants pull towards light, so if you start to see your plant leaning too far to one side, almost ready to topple over, that's the plant telling you to rotate it. Rotating your plant can help its growth become more balanced.

At the end of the day, just be aware that your plants are talking to you. If you see a flower bloom or a new large leaf unfurl, that's a happy plant. It's telling you thanks for providing it with the sun and water it needed. You feel rewarded as a plant parent. Way to go, you! Now go treat yo'self to another plant. You deserve it!

THE ANATOMY OF A DIAMOND

Have you ever heard the old adage, "Every time a fiddle-leaf fig unfurls a new leaf, an angel gets its wings"? Yeah, me neither. But I would totally believe it if I wasn't the one that just thought of it and wrote it here. I love thinking about the beauty that greenery creates. From time to time, I find myself thinking about my life and how I lived it before deciding to bring plants into my home. I call this time B.P. (Before Plants). You see, before plants, I was filled with stress and concern, not really settled or satisfied with the path I found myself treading. I felt alone going down a road that I prayed would lead to a fork, rather than to an end. To the outside world I might have seemed well put together, but I was never whole. It all changed for me once I started to bring plants into my home. I know that sounds far-fetched, but I have no other explanation for the change in me and the things around me, A.P. (After Plants). Bringing plants into my home changed the overall look from stale to alive, changed the mood from dark to vibrant, and turned every hard line into a soft stroke, but it's in my education on plant care that I have discovered my true self.

It's the lessons learned from tending to a plant's needs that have helped me become more aware of my own and of those around me. For instance, when watering a plant, it's so important to pay close attention to the moisture level of the soil and allow that to dictate when you give it a drink. If you're able to make yourself aware of these small shifts and nuanced changes, you can fix any problems before they become real issues. This is where you start to see your plants thrive, and if you're able to apply that knowledge to other relationships in your life, you'll see them thrive as well. The beauty in that vibrates.

I learned to start with myself. To pay close attention to my needs, listen, and water properly. I can't tend to the needs of others if I haven't made myself right first. That's why, on an airplane, they tell you to put your own oxygen mask on before you assist anyone else with theirs. There are immense similarities between plant care and self-care. One could even say that plant care is self-care.

LOVE VIBE

My wife Fiona and I moved into this apartment because of all of the light it had, knowing this would be an environment which would help our plants to thrive. We've watched our love for each other thrive right alongside them.

THE WALLS ARE ALIVE
Having a living wall indoors is the most creative way to blur that line between indoors and outdoors. A work of living art will make a home feel lush, vibrant, and alive.

In caring for my plants, I've found an inner peace that wasn't there before. It's as if a precious jewel has been unearthed from within me and now I get to walk around gazing through its crown. Yeah, it's really like that. I can honestly say I've been reaping the real, tangible benefits of having houseplants.

I see this mostly reflected in my relationship with my wife, Fiona. With her being the most important thing in my life, seeing her thrive is essential. In past relationships, before having plants, I would let the stress of the world, my job, and my insecurities paint a dark cloud around me. This ultimately withered the life out of those relationships. Unknowingly, as I started to bring greenery into my home, I became less stressed, less anxious, and with that, more attentive, more relaxed, more confident, and eventually a better version of myself. While I can say maturity played a large role in this shift, it was finding myself submerged in plant care that helped hone this.

While having plants in your space can make you feel creative and experience a sense of euphoria, it's in caring for and connecting with your plants that you truly see this therapy take shape. You find yourself deeply immersed in your routine, leaving all of the stresses of the world behind you for the moment. There aren't many things that we do nowadays that don't involve multitasking. In such a "go, go, go" society, proper plant care forces us to slow down, see the needs or issues, and focus on care. For me, when I water my plants, I do so with intention. I start by making sure that the water being poured into the watering can is lukewarm by testing it as the can fills, as a parent would test the temperature of the milk in a baby's bottle. I then take that can, and as if using the pour-over coffee method, I pour the water slowly around the top of pot, saturating the topsoil of the plant to ensure that every root below has the opportunity to grab on to a bit of moisture before it makes its way down the pot and out of the drainage hole. Interesting side note here, in the coffee world, when they use this method

with coffee grounds, it's called the bloom pour. I love that. When I'm wiping down foliage to remove dust or to check for unwanted critters, or when rotating the plant to make sure that each part of it has its time in the sun, I speak gently to the plant as if talking to a friend or pet that I care for. This connection increases my determination to make sure the plant thrives.

All of these subtle care techniques I've weaved into my relationships with family, friends, and those around me. There are times when the world calls for you to be fully in "go, go, go" mode but you have to find the time to slow it down, focus on the ones you love, and care for them with intention. In doing so, maybe you'll unearth that jewel within you, and find that happiness in plants everyone talks about. I hope it's a diamond.

WILD PLANTS

So many plants, so little time. So little space as well, right? With each book I'm fortunate enough to create, I feel it's only right to share some of my favorite plants that I have brought into my home. And believe me, I've brought in many. At this current moment I have a little over 200 in my apartment and a few more in my workspace. Please understand that that is a mix of all different types and sizes of plants, and 66 of the 200-plus plants I care for are just cuttings that are displayed in my propagation wall. I know my limit and where I am now, is it. OK, if someone decided to build me a greenhouse I guess I'd have more plants. I'd be crazy not to. With the light that I have in my home and workspace, and the amount of time I have available to care for them, I've been particular with the selection of plants that I bring in. So when it comes to sharing about these wonderful houseplants, and hopefully inspiring you, the reader, to bring them in, I want to be considerate of the fact that we don't all have the same amount of light coming into our homes, the same amount of time on our hands, and the same type of knowledge when it comes to care level. So here, I've selected plants that can work in low-light to bright light spaces, and those that are more low-maintenance to aggressively needy. I've mixed in a little jungle with a little desert but made sure in the end that, whatever you decide to bring into your space, it will help that spot feel more warm, lush, and alive. And please note: the information provided about the light the plants discussed here need is based on those in the northern hemisphere. If you live in the southern hemisphere, you will need to reverse the direction given.

AIR PLANTS

(*TILLANDSIA*)

There are few plants as versatile and unique as the family of *Tillandsia* or, as they are better known, air plants. There are over 650 beautiful species in the *Tillandsia* genus and if you're fortunate to travel around the deserts and jungles of northern Mexico, you might get to see some of them growing in the wild. But you'll have to look up from the surface of the ground to do so because air plants don't grow in soil, they actually grow on other plants. Just like staghorn ferns (*Platycerium bifurcatum*) or orchids, air plants are epiphytes, which means they are plants that grow on other plants, using them as a host. In their natural habitats you'll find them clinging to shrubs, bushes, and trees. While on a trip to Tulum, Mexico, with my wife, we came across many *T. stricta* air plants growing on the trees above. I recall the incredible urge I had to climb those trees and pull down as many as I could. The charm of these plants is just so alluring. Knowing that those air plants were better off living their best lives in the jungles of Tulum, I suppressed my urges and didn't disturb the plants. OK, let's be honest, the main reason I didn't is because it's illegal to remove greenery from those jungles and even if I could, you're not allowed to bring them back into the United States. So when I saw air plants like Spanish moss (*T. usneoides*) growing on the trees in Houston, Texas, or New Orleans, Louisiana, it was time for me to bring a little of that dripping moss into my home.

When it comes to bringing air plants indoors, this is where they separate themselves from other houseplants. Because they don't require soil to grow, this means they can be displayed almost anywhere in the home. That word "anywhere" is used lightly here because you'll still need to find the right light for them, but you're able to get creative with the way you style them. I've draped Spanish moss across wooden hangers. I've placed *Streptophylla* air plants in bowls and displayed them on tables. If you look at *The Air Plant Wreath* project on p.103, you'll see how the potential for styling with air plants is almost endless. But there are many things to know and understand about their needs before you can just go randomly placing them in your home. Knowing how to properly care for them will help you see their true bloom right in front of you. So here are my tips on how to care for them.

AIR PLANT TYPES SHOWN:

1 ***TECTORUM***
2 ***SELERIANA***
3 ***STRICTA***
4 ***STRICTA***
5 ***XEROGRAPHICA***
6 ***STRICTA***

LIGHT

When it comes to light, style your air plants in a spot that gets anything from bright indirect light to indirect light. Given that they grow on trees and bushes in the wild, this means they're under a canopy of foliage getting dappled light, but just because they might be in a shadier spot outdoors doesn't equate to them loving a shadier spot in your home. The brighter, the better. But you'll want to refrain from placing them in direct sun unless you have a *Streptophylla* air plant, which can tolerate a little direct sun. For all other air plants, direct sun will cause them to wilt and dry up quickly, making it more difficult for you to give your plant the moisture it needs. If you have northern or eastern exposure, this would be the perfect spot for plants like this.

WATER

Speaking of moisture, contrary to popular belief, air plants indoors aren't just going to get their fix of water from the air. In their native environments, where it's humid and pours when it rains, yes. Humidity and moisture are everything and getting that right indoors can be tricky. To keep your air plants thriving indoors, once a week during the warmer months and once every two weeks during the colder months, you'll want to take them from wherever you have them displayed and submerge them in a sink or tub full of lukewarm water. Let them bathe in the water for about 15–30 minutes. Submerging is the best way to make sure all parts of the plant get the moisture they need. Once they have had a good soak, take them out and let them drain on a drying rack, cup-side down, before placing them back in their spots. The reason for drying them cup-side down is to ensure that all of the water drains from the interior of the plant. If you let water stay inside the base of the air plant it'll cause the roots to rot, the base of your air plant will turn dark brown and soft, and slowly the plant's foliage will fall apart. Once they've had time to dry, just give them one quick shake to remove any remaining droplets of water that could still be tucked inside their foliage, before putting them back.

To make sure they are getting the humidity they need, mist them with lukewarm water at least twice a week during the warmer months and once a week during the cooler times of the year.

Display your air plants with the cup of the air plant pointing down, so that when you mist them, the moisture falls out of the cup. Mist that pools at the base of an air plant could lead to it rotting.

PROPAGATING

Propagate your air plants using the division and separation method. Air plants will grow new pups on their sides as the plant matures. Once a new pup is at least one-third of the size of the mother plant, it can be separated from the mother by gently pulling it at its base.

BLOOMING

When you're giving them everything they need, air plants will reward you by producing a beautiful flower. But understand that this a sign of the plant's maturity which you'll only see once. And once they flower, the plant's life is nearing an end. Their blooms can be in many vibrant colors, like red or purple. Enjoy them while they last because these flowers will live for about a week and then die off.

TROUBLESHOOTING

SOGGY LEAVES This is a clear sign that the plant is being overwatered. You have to really try hard to overwater an air plant, so remember to mist them more than bathe them.

BROWNING OR WILTING

This is a sign of an underwatered or dry plant. This could be due to the fact that the plant is getting too much direct sun or it's just not being watered as consistently as it should. Misting air plants throughout the week will help this. Also, never place them near air vents or heaters, as this will cause them to dry out as well.

STYLING TIP

Why is the Christmas tree the only tree that gets to be decorated? Why not hang some air plants from the branches of your larger indoor trees?!

ALOCASIA 'PORTODORA'

(*ELEPHANT EAR*)

Have you ever heard someone say, "If you love it so much, why don't you marry it?" Well, the way I feel about the *Alocasia* 'Portodora', I may need to get down on one knee. Look, I'm already married but if I wasn't... OK, that's taking it a bit too far, but I really do love this plant. I'm sure anyone with eyeballs would quickly see the allure. With its large, green arrowhead foliage that points straight up toward the sky, as if opening its arms to embrace the kiss of the day's light, when brought indoors, it's an instant eye-catcher. Known by many as the elephant ear plant, because of the ruffled, floppy shape and the large size of its leaves, it has a demanding presence, just like an elephant. When given the proper care, you could see yours grow to about 5ft (1.5m) indoors, so make sure to give it room to show off. Even when unfurling new growth from its center, it has an elegance which resembles that of a blooming flower.

Like the *Monstera* or bird of paradise (*Strelitzia reginae*), the *A.* 'Portodora' is a tropical plant that can help seamlessly blur the line of indoor/outdoor when styled in a home. For those with outdoor space, giving it a little vacation outdoors during the spring and summer will help provide it with the climate it is used to in its native environment. And, when bringing it indoors, it's important to try your very best to mimic that environment so that your plant can continue to thrive. So here are some tips on how to best care for your *A.* 'Portodora'.

LIGHT

For all plants, light is everything. With the *A.* 'Portodora', it requires the best light out there, and that's bright indirect light. If you have medium/indirect light, it will tolerate that, but it must stay away from direct sun. While direct sun in the morning might not be as intense, direct sun in the afternoon will burn the foliage and kill your plant over time. So placing it in an eastern- or southern-facing window would be perfect. Understand that while it likes dappled or indirect light, it won't survive in low light. If you find yourself hell-bent on bringing an *A.* 'Portodora' into a spot that get direct sun, please protect it with some sort of filter, like blinds or white curtains, so that the direct sun can be diffused.

WATER

Alocasias are tropical plants, so you'll want to make sure they have the right level of humidity. Getting this right indoors requires a bit of work. Just like a fern, to ensure your *A.* 'Portodora' is living its best life, you have to be extremely attentive, meaning you'll need to look over the plant daily. For this plant, the soil must be kept evenly moist. That is key. And when we say evenly moist, know that that's different to wet. Prolonged wet soil will cause many problems, so to avoid this, always check the soil by using your finger or a moisture meter. Sticking my finger into the soil by about 2in (5cm) will give me a perfect understanding of the moisture level of the soil. If the top 2in (5cm) are a little dry, it's time for a drink. Make sure to use lukewarm water and slowly work your way around the pot, watering the soil as you would a cup of pour-over coffee. To help retain the humidity that this plant needs, misting it every morning is advised. Again with lukewarm water, mist the bottom sides of the foliage so that the mist trickles down the stem of the plant and doesn't pool on the top of the foliage. Water resting on top of the leaves can lead to a bacterial infection. Like most plants, during the colder months of the year, you'll want to cut back on the watering for two reasons: one, because some of your houseplants will go dormant, and two, because the soil won't dry up as fast as it did when it was warmer. Again, checking the moisture level of the soil every time before watering will help you keep your plant healthy. One of the ways I like to maintain moisture in the soil is to insert a water spike in the soil and place a wine bottle filled with water on top (see *Watering Your Plants While Away* p.112). This will allow moisture to be slowly released throughout the week.

SOIL

As mentioned, you'll want to keep the soil moist but also well-drained, so having additives in your potting mix that can help retain moisture is key (see *Creating Your Own Potting Mix* p.152). Adding peat moss to the soil can help hold the moisture in. You'll also want the soil to be rich in nutrients, so having an organic, breathable soil full of peat moss and pine bark is important.

REPOTTING

The sign that it's time for your plant to go into a larger pot is when you start to see its roots emerging from the drainage hole. When it's time, grab your spade or trowel and insert it between the pot and the soil. Do this around the entire pot. Once it's loosened, gently wrap your arms around theplant to support the foliage, and slowly lift it out of the pot. Make sure the new pot you're placing it in is 2in (5cm) larger in diameter than the previous pot. And again, make sure that the new pot has proper drainage.

TROUBLESHOOTING

DROOPY LEAVES If you start to see the leaves faint or droop, this is a sign that the plant isn't getting enough bright light or is too wet. Indoors the foliage won't always be pointed straight up because your plant will want to pull toward the light, meaning that it might lean in the direction of the nearest window and the sky. But if the arrowheads start pointing towards the floor, you have a huge problem.

YELLOWING LEAVES If you start to see the leaves turning yellow, that is a sign of too much water. Remember to keep the soil evenly moist and not wet. If you see yellowing happening, check the base of the pot and make sure your plant is draining well. In cases like this, I'll take a chopstick and poke holes in the soil to aerate it a bit and give it a better path for watering.

WEBS AROUND LEAVES
Alocasias are notorious for attracting spider mites. These bugs will create webbings around the foliage and kill your plant if not taken care of. Always check around the edges and beneath the foliage for spider mites. Getting rid of them is fairly easy. Take a damp cloth with a dab of

mild dish soap and wipe down the leaves. You'll probably want to do this once a week until the issue ceases.

STYLING TIPS

Place this beauty in a glazed ceramic container, under a skylight if you have one. To keep this plant looking its best, remove any dried leaves with a pair of sharp shears. I suggest choosing a spot for this plant away from smaller plants. Because of its large foliage, it will cast a heavy shadow on many of the plants below and around it.

BLUE AGAVE

(AGAVE TEQUILANA)

When is it not a good time to spice up your space with plants? I mean really, is there a time that you couldn't bring a plant into your home? I think not. And one of the plants that I have in my home and love bringing into other indoor spaces is the amazing blue agave, better known as the *Agave azul*. This beautiful Mexican desert plant is coined the blue agave because of the hints of blue coloring in the foliage. And its foliage comes bearing a bark and a bite. Like long swords sticking out of the soil, the tip of each leaf is sharply pointed and can harm you if you're not careful. But don't blame the plant, it's just trying to protect itself. So if you have small kids or pets, it's important to place your agave plant in an area of your home that is out of reach.

LIGHT

When it comes to light, always consider how your plant grows in its natural environment. For the agave plant, it grows in a desert with lots of direct sunlight. So in your home, you'll want to place it in a spot that gets at least 4–6 hours of direct sunlight a day. For those with west- or south-facing windows, this is where you'll want to give it a home. Just like your other succulents, having a lot of direct sun will be best for it. There's no low light in the desert, so don't try to force it into the low-light corners of your home. Remember, we are dealing with a living thing and in order for it to live, it needs the proper amount of light and care.

If the temperature is right during the spring and summer where you live, transporting it outdoors would be perfect and it will thank you for it by growing, but doing so very slowly. So don't be offended if your plant hasn't grown much over the years. The agave is slow-growing and doesn't require too much of your attention. Just give it the full sun that it needs and let it be. If you're someone that typically likes to helicopter parent your plants, this might not be the plant for you.

WATER

Water the agave once the top half of the soil is completely dry, using lukewarm water. During spring and summer this could be once a week but in winter, once a month. This will also depend on if it lives outdoors or in. If outdoors, its soil will dry out much faster, so for a younger plant, you'll probably be watering it once every five days, and as it matures, once a week. It is drought-tolerant, so that helps. If indoors, you'll water less. But let the moisture level of the soil guide you before giving this beauty a drink. And make sure to only use lukewarm water. Pour the water slowly and evenly across the top of the soil, making sure that all sides of the pot and the roots below can have a drink. You should continue to water your plant until you see water exiting out of the drainage hole and into the base tray. Once that happens, let the runoff water sit in the base tray for about 15–30 minutes, just in case the roots and soil didn't get a chance to hold on to that water as it came rushing down. After 30 minutes, if there is still water in the base tray, take your plant off that base tray, dispose of that runoff water, and then place your plant back in the tray.

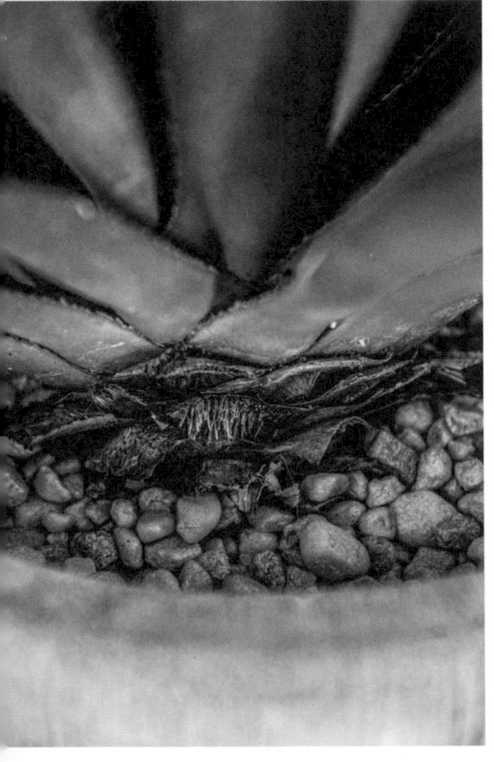

SOIL

The soil of your agave needs to be fast-drying and well-aerated. Having additives like sand or perlite in your soil will help that process (see *Creating Your Own Potting Mix* p.152). To help aerate the soil, poking a wooden dowel or chopstick down into the soil can help. So that your soil dries properly, I recommended dressing this plant in a porous container, like a clay or terracotta pot.

REPOTTING

The blue agave, along with the others in the agave genus won't require repotting often. Because it is such a slow-grower, its roots will take time to push themselves out of the drainage hole, warning you to repot them. Make sure to repot in a planter that is 2in (5cm) larger in diameter than the previous pot. Consider the material the pot is made out of because this will be a factor in the moisture level of your soil. Repot your plants during spring and summer for the best results.

PROPAGATION

When it comes to propagation, use the division and separation method here by separating any new pups that grow out of the soil from the roots and planting them in a new pot.

TROUBLESHOOTING

DRIED LEAF TIPS This is a sign that your plant isn't getting enough water. While you want to be cautious not to overwater it, on the days that you do, it's important to make sure to water it all the way through.

SOGGY OR MUSHY LEAVES

This can mean that you've overwatered your agave and it possibly has rotting roots. Carefully remove any soggy leaves with a sharp pair of shears and check the soil and roots of your plant. The roots should look off-white and sturdy, like an *al dente* noodle. If any look dark brown and mushy, that is a rotting root. Cut them back with shears and repot in fresh, fast-drying soil.

STYLING TIP

I find this plant looks more natural in a terracotta or clay pot and placed on a plant stand or pedestal. With the blue agave being the main ingredient in tequila, placing one near your bar could really make that area of your home come to life.

KANGAROO FOOT FERN

(*MICROSORUM DIVERSIFOLIUM*)

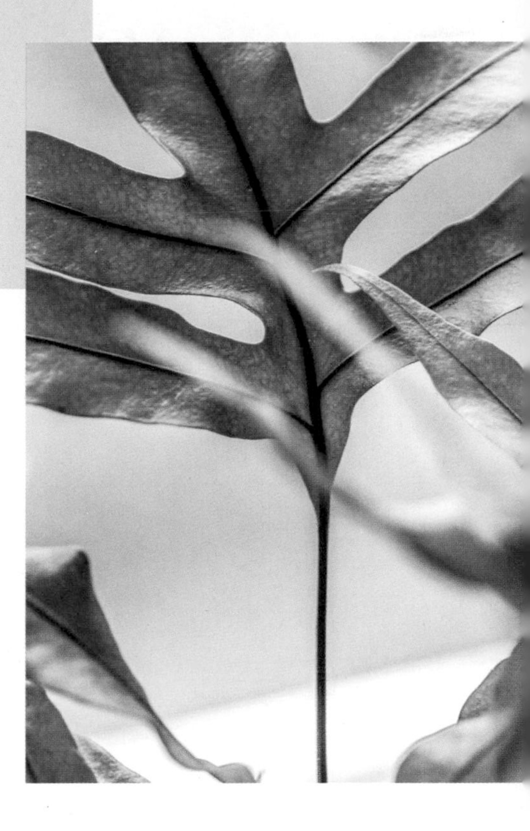

The first time I laid eyes on a kangaroo foot fern I knew I needed one as a part of my collection. While I'm self-aware enough to know I shouldn't have many ferns in my home due to the fact that I travel often, the kangaroo foot fern is one of the more forgiving in the family of ferns. In being more forgiving, it's also less fragile than other ferns in the world. Named after the foot of the animal that shares its native land, when looking at the shape of the fronds of the kangaroo foot fern, it's clear why. And because I'm always down to name my plants, I named my first kangaroo foot fern Joey. No judgments, please. With its thick, waxy green fronds and black stems that unfurl from its hairy, caterpillar-like rhizomes, this plant has all the attitude to spice up your collection. Here are my tips on how to care for the kangaroo foot fern.

LIGHT

When it comes to light, treat your kangaroo foot fern like most ferns, and style it in a spot that gets anything from bright indirect light to indirect light. While it will tolerate medium light, the more you can expose it to open sky, the better. Dappled light is what it's used to in the wild, so avoid placing your fern in direct sun. Direct sun will burn the fronds of your fern, causing them to wilt and dry up quickly, making it more difficult for you to give your plant the moisture it needs. If you have northern or eastern exposure, this would be the perfect spot for plants like this.

WATER

The kangaroo foot fern is a tropical plant, and like all tropical plants, it requires humidity to thrive. Humidity and moisture are everything, so you'll want to mimic a tropical environment in your home when caring for this fern. This all first starts with how you water it. Make sure to water your kangaroo foot fern so that its soil stays evenly moist, not wet. Like many of our plant friends, they are sensitive to being overwatered, so check the soil with your finger, a wooden dowel, or a moisture meter. If you're someone who is considering adding a lot of ferns to your collection, a moisture meter will be helpful. When it's time to water, only use lukewarm water. Pour the water slowly and evenly across the top of the soil, making sure that all sides of the pot and the roots below can have a drink. You should continue to water your plant until you see water exiting out of the drainage hole and into the base tray. If the plant is in a hanging pot, allow it to drain completely before placing it back in its spot.

Throughout the week, to increase the humidity around

7 27
Renoir
JOHN ELDERFIELD the CUT-OUTS of HENRI MATISSE George Braziller
PICKVANCE VAN GOGH IN ARLES MMA / ABRAMS
GASTON DIEHL · THE MODERNS

your fern, you'll want to mist it with lukewarm water, making sure to get underneath the fronds so that the moisture falls down the plant and into its roots. Mist that builds up on the surface of the frond could lead to bacteria building up.

SOIL

To keep the soil moist and well-drained, having additives in your potting mix that can help retain moisture is key. Adding peat moss or vermiculite to the soil can help hold the moisture in. You'll also want the soil to be nutrient-rich, so having an organic, breathable soil full of peat moss and pine bark is important (see *Creating Your Own Potting Mix* p.152).

REPOTTING

Repot this plant during spring and summer only. While it is OK with being in tighter spaces, once you see roots emerging from the holes of the base tray, this is a clear indicator that it needs a larger pot. Make sure that its new pot is only 2in (5cm) larger in diameter than its previous pot.

TROUBLESHOOTING

YELLOWING FRONDS This is a clear sign that the plant is being overwatered. A good rule of "green" thumb when it comes to the kangaroo foot fern is to watch the stiffness of the antler fronds. If the fronds are upright and stiff, it's well-watered. If they start to droop a bit, it's time for a drink.

BROWNING TIPS OR CURLING FRONDS This is a sign of an underwatered or dry plant. This could be due to the fact that it's getting too much direct sun or it's just not being watered as consistently as it should. To avoid this, make sure you're watering it consistently and keeping the plant away from direct sun.

STYLING TIP

Let your kangaroo foot fern shine to the best of its ability by placing it in a hanging planter so that it can cascade out of the pot and show off its personality. To help with moisture, plant it in a glazed ceramic planter or plastic pot. Remember to rotate your fern at least once a month for balanced growth.

MAIDENHAIR FERN

(*ADIANTUM RADDIANUM*)

When it comes to the maidenhair fern, many plant parents find themselves having difficulties keeping them alive indoors. But because of how stunning the maidenhair fern is, plant lovers just can't help bringing one home and trying over and over again. With its elegant, green, animal-paw-like fronds, which unfurl from its thin black stems, it's easy to see the attraction. Believe me, I totally get it. But let's remember, we are dealing with a living thing, so knowing how to properly care for it before bringing it into your space is helpful. This fern is just one of the 200-plus beauties in the *Adiantum* family that grow around the world, and I can admit that I've never met a fern I didn't instantly want to bring into my home. The maidenhair fern is no exception.

LIGHT

When it comes to light, give your fern what it needs or you'll find yourself with a dead plant in no time. Like most ferns, you want to dedicate a spot for it that gets anything from bright indirect light to medium light. Yes, I say that a lot but, again, all plants would love this type of light. The more you can expose it to open sky, without having it kissed by the sun, the better. If you have medium light, they'll tolerate that as well. But you must stay away from direct sun. Remember, think about where you see ferns growing in nature. Low on the Earth's surface, dancing in dappled light. Yes, you might find them getting a little direct sun in the wild, but once brought indoors, sunlight through glass is a different type of beast. When direct sun makes its way through glass, it makes the light more intense and will cause the fronds of your fern to wilt and the soil to dry up quickly, making it more difficult for you to give your plant the moisture it needs. So knowing that the more it's exposed to brighter light, the faster your soil will dry up, don't be surprised if you find yourself watering more often during the warmer months of the year.

WATER

Speaking of water, just like most ferns, you want to make sure it has the right level of humidity. For the maidenhair fern, moisture is everything and getting that right indoors can be tricky. To keep this fern thriving, you'll have to be extremely attentive, meaning you'll have to look over this plant daily. This brings me back to my section on *Cat People vs Dog People* (see p.178). The maidenhair fern is basically like having a puppy. Your job is to make sure you are keeping the soil evenly moist. That is key. When I say evenly moist, know that that's different to wet. When it comes to differentiating between the two, for evenly moist, I like to use the finger test. If you press your finger down onto the soil and a little dirt or mud sticks to your finger when you remove it, that's evenly moist. But if you press your finger down on the soil and water pools at the side of your finger, that's wet. And wet is not what you want. This probably means your pot isn't draining properly and you'll want to remedy that right away. To help give this fern the humidity it needs, you'll want to mist it every morning, with lukewarm water. This means you'll spend some time checking the soil day to day. That's why it's important to be aware of the type of person you are. If you travel a lot or are a bit forgetful, this plant probably isn't the one for you. For ferns, having the help of a moisture meter can make a huge difference.

SOIL

As mentioned, you'll want to keep the soil moist but also well-drained, so having additives in your potting mix that can help retain moisture is key. Adding peat moss to the soil can help hold the moisture in. You'll also want the soil to be nutrient-rich, so having an organic, breathable soil full of peat moss and pine bark is important (see *Creating Your Own Potting Mix* p.152).

REPOTTING

The maidenhair fern is a slow-grower, so you probably won't need to repot as often as you will your other plant friends. A sign that it's time for it to go into a larger pot is when you start to see its roots emerging from the drainage hole of the pot. While it will manage with its roots being a bit crowded, moving it into a larger pot at this time would be ideal. When it's time, grab your spade or trowel and insert it between the pot and the soil. Do this around the entire pot.

Once it's loosened, with one hand gently supporting the foliage, use the other hand to turn the pot over. Here, you'll have an easier job of removing the plant from the pot. With your hand supporting the foliage, grasp the base of the stems and gently pull the plant from its pot. Make sure the new pot you're placing it in is 2in (5cm) in diameter larger than the previous pot. And again, make sure that the new pot has proper drainage.

TROUBLESHOOTING

SPORES If you see these little bumps on the underside of the leaves, don't freak out. Your fern isn't infested with bugs. These bumps are spores that appear when the plant is mature and ready to reproduce. You probably won't have success trying to propagate using these spores but do know that if your maidenhair fern has them, it's a happy plant!

YELLOWING LEAVES When you start to see the leaves turning yellow, that is always a sign of too much water. Remember to keep the soil evenly moist and not wet. If you see yellowing happening, check the base of pot and make sure your plant is draining well. In cases like this, I'll take a chopstick and poke holes in the soil to aerate it a bit and give it a better path for watering.

WILTING LEAVES When the leaves start to curl and wilt, that's a clear sign that your fern is in desperate need of water. That means drop everything and water this plant. If a frond or an entire stem dies on you, don't panic. If given what it needs, new growth will occur.

STYLING TIPS

Place this beauty inside a terrarium to show it off, or on a table next to other ferns. To keep it looking its best, remove any dried leaves with a pair of sharp shears. Be careful and precise here, because you don't want to accidently cut off a healthy stem.

PONYTAIL PALM
(*BEAUCARNEA RECURVATA*)

Calling this plant a palm can be quite deceiving to those thinking about the care that other palms out there need and that are looking to do the same with their ponytail palm. This is because the ponytail palm isn't an actual palm. It's more like a succulent. So when caring for it, you should consider the conditions of its natural habitat, Mexico. Some know it as the ponytail palm, which comes from the fact that its foliage grows out like a ponytail protruding from one's head, while some call it the elephant's foot plant due to the fact that its trunk (no pun intended) resembles an elephant's foot. What I adore about this plant is how large it can grow (15ft/4.5m), how long it can live (300 years), and how easy it can be to provide the care that it requires. I'm a huge fan of a plant that wants me to love it but doesn't want me to coddle it. I'm already busy giving all of that attention to my Australian tree fern (*Dicksonia antarctica*). When it comes to the ponytail palm, it's best to just let it be and tend to it when it calls on you. Here are my tips on how to care for your ponytail palm.

LIGHT
When it comes to light, make sure to dedicate a spot for it that gets anything from bright indirect light to direct sun. The brighter, the better. The closer you can mimic the natural environment for this plant, the more you'll see it thrive. This means if you have any south- or west-facing windows, those would be the perfect spots for your ponytail palm. Try your best to avoid low light. Those dark corners of your home can eventually kill this plant. Because the ponytail palm grows in a bush shape, it is important for you to rotate this plant often for balanced growth.

WATER
Just like most desert plants, the ponytail palm wants to dry out completely between waterings. For this plant, I say it's better to lean towards underwatering than overwatering. If you're ever unsure, just give it a few more days and then come back to it. Ponytail palms are susceptible to overwatering, which will lead to root rot and a dead plant.

Let's stay away from that. So as I've said before, let the moisture of the soil tell you when the plant should get a drink. Since this plant wants its soil to be completely dry, let it dry out. Remember when we discussed the elephant foot-like trunk? Well, what that trunk does is store water for the plant, just like a cactus, to get through the dry times in the desert. So if you feel like you missed a watering, don't stress about it. Again, always use lukewarm water and when you do so, pour it slowly over the top of the soil, like pour-over coffee. Allow that water to trickle down through the soil, giving each root the opportunity to pull moisture in. Once that water starts to seep out of the drainage hole, you've given your plant enough to drink. Let that runoff water sit in the base tray for about 30 minutes but then after that, dispose of any leftover water. Never, ever let your plants sit in water.

SOIL

For your ponytail palm, you'll want to use a potting mix that is well-aerated and dries out fast. In this case, having additives in your potting mix that can help pull moisture away from the soil so that the roots aren't sitting in water for long is important (see also *Creating Your Own Potting Mix* p.152). Depending on the planter, I like to create a potting mix that is 60% soil, 20% perlite, and 20% sand for this plant. This would be my ratio if I was placing the plant in a clay or terracotta pot, but if it was going in a glazed ceramic pot or plastic container, I would add more perlite.

REPOTTING

Repot your ponytail palm only when the roots of the plant start making their way out of the drainage hole. The more the roots grow at the base of the pot, the more susceptible they will be to rotting. Make sure to repot in a planter that is 2in (5cm) larger in diameter than the previous pot. Consider the material the pot is made out of because this will be a factor in the moisture level of the soil. The ponytail palm is a slow-grower, so you won't be repotting often. Repot your plants during spring and summer for the best results.

TROUBLESHOOTING

YELLOWING LEAVES This is a sign that your plant is being overwatered. Remember, it doesn't take much to overwater a ponytail palm, so let it dry out completely.

BROWN TIPS OR CRACKS ON THE LEAVES This is a sign of underwatering. While you want the plant to dry out completely, it can't go months without water. Check the soil to see if it's dry. If the plant is small enough, pick it up and check the soil around the drainage hole. If it's dry, water it. If the plant is too large to pick up, take a wooden dowel or chopstick and stick it down deep into the soil. If it comes up dry, it's time for a drink. If it comes up a bit muddy, let the soil dry out.

BROWN SPOTS IN THE INTERIOR OF THE LEAVES This is a sign of root rot. Take your plant out of its pot and check the soil and roots. The roots should look off-white and sturdy, like an *al dente* noodle. If any look dark brown and mushy, that is a rotting root. Cut them back with your shears and repot in fresh, fast-drying soil.

STYLING TIP

With the ponytail palm wanting to dry out completely between waterings, this gives you the ability to style the topsoil with stones, rocks, marbles, etc., because you won't need to stick your finger in once a week. To keep the plant looking its best, remove any dried leaves using a sharp, clean pair of shears and when they need a little dusting, take a duster and wipe the leaves. Removing the dust that has accumulated on the foliage will bring more light to the tissue of the foliage and give your plant back its natural shine.

PHILODENDRON 'ROJO CONGO'

Let's just get right to it. I... love... this... plant! In my home we have this plant styled over our bed, hanging from a plant hammock my wife made. Yes, I said plant hammock. You can find out how to make one for your own plant in my first book, *Wild at Home*. But enough about me. Let's discuss the charm of the 'Rojo Congo' philodendron. After previously having a macho fern over our bed and all the work that was required to keep it moist, I decided we needed a less fussy plant. In comes the 'Rojo Congo', whom we've named Big Red. While still adding that sense of sleeping in a jungle under the large foliage which creates a canopy of greenery, it never demands too much attention. Well, that's only when we're talking about care; it definitely demands your attention when it comes to its look. It gets the name 'Rojo' because of how red the new growth is as it unfurls from the single stem of the plant, but, as the leaves mature, they start to take on a more of a deep green hue, with a red trim around the edge, this all being held by an elegant burgundy stalk. And the charm of this plant doesn't stop there. When given the right care, you'll be treated with not only growth, but also a hearty red flower that blooms from its stalk. Now this is one sexy plant! Here are my tips on how to care for your 'Rojo Congo'.

LIGHT

When it comes to light, make sure to dedicate a spot for this plant that gets anything from bright indirect light to medium light. If you are going to place it in direct sun, make sure it's from an eastern-facing window, not western. AM direct sun isn't as harsh as PM direct sun, and too much direct sun can burn the foliage, causing it to develop brownish-orange spots, and can eventually kill the plant. Find a spot that has a lot of exposure to open sky or a room that filters direct sun into bright indirect light.

WATER

Like all philodendrons, you want to make sure this one isn't overwatered. The best way to keep the 'Rojo Congo' healthy and thriving would be to water when the top 2in (5cm) of the soil are completely dry. To check this, I recommend the finger test. Stick your finger into the soil about 2in (5cm) and if it comes up muddy, the plant is not thirsty, but if it comes up dry, then it's time to give it a drink. If you're a plant hoarder like me, sticking your fingers in soil every day/week can become a bit much, so you might want to invest in a moisture meter to help you with the heavy lifting.

Make sure to only water with lukewarm water. Pour the water slowly and evenly across the top of the soil, making sure that all sides of the pot and the roots below can have a drink. You should continue to water your plant until you see water exiting out of the drainage hole and into the base tray. Once that happens, let the runoff water sit in the base tray for about 15–30 minutes, just in case the roots and soil didn't get a chance to hold on to that water as it came rushing down. After 30 minutes, if there is still water in the base tray, take your plant off that base tray, dispose of that runoff water, and then place your plant back in the tray. Given that it's a tropical plant, you'll want to create the climate in your home that it can best thrive in. This means bringing in some humidity. For the 'Rojo Congo', during the warmer months I recommend misting it at least twice a week and during the colder months, you'll want to mist it every morning with lukewarm water.

SOIL

For your 'Rojo Congo' and other philodendrons, use a potting mix that is well-aerated and dries appropriately. In this case, having additives in your potting mix that can help pull moisture away from the soil so that the roots aren't sitting in water for long is important. Depending on the planter, I like to create a potting mix that is 80% soil and 20% perlite for my philodendron. This would be my ratio if I was placing the plant in a clay or terracotta pot, but if it was going in a glazed ceramic pot or plastic container, I would add more perlite or even sand (see *Creating Your Own Potting Mix*, p.152).

REPOTTING

Repot your 'Rojo Congo' only when the roots of the plant start making their way out of the drainage hole. It's always the root growth that determines when you should repot, not the size of the plant. The more the roots grow at the base of the pot, the more susceptible they will be to rotting. Make sure to repot in a planter that is 2in (5cm) larger in diameter than the previous pot. Consider the material the pot is made out of because this will be a factor in the moisture level of the soil. Repot your plants during spring and summer for the best results.

PROPAGATION

For the 'Rojo Congo' and most philodendrons, propagate using the stem-cut method. This means locating the node of the stem, cutting below that node, and submerging that node in water. Over time that node will start to develop roots and once those roots are about 4–6in (10–15cm) in length, you can then place the cutting in soil. Propagate your plants during spring and summer for the best results.

TROUBLESHOOTING

YELLOWING LEAVES This is a sign that your plant is being overwatered. If you've been following the tips on how to water but are still seeing yellowing leaves, you should use a wooden dowel or chopstick to aerate the soil and if that doesn't help, pull the plant out of the pot and check the soil. If the soil is staying wet for too long, this could be the reason.

BROWN TIPS OR CRACKS ON THE LEAVES This is a sign of underwatering. Remember to check the soil and make sure that when you do water your plant, that water runs through all of the soil. This could also be a sign of too much fertilizer.

BROWN SPOTS IN THE INTERIOR OF THE LEAVES This is a sign of root rot. Take your plant out of its pot and check the soil and roots. The roots should look off-white and sturdy, like an *al dente* noodle. If any look dark brown and mushy, that is a rotting root. Cut them back with your shears and repot in fresh, fast-drying soil.

RED SPOTS IN THE INTERIOR OF THE LEAVES This could be a sign of a bacterial infection. It could be caused by too much moisture lingering on the top of the foliage, so when misting, make sure to mist the underside of the leaves.

STYLING TIP

Unlike most philodendrons, the 'Rojo Congo' isn't a climbing plant, which means it doesn't use its aerial roots to climb plants, rocks, and other structures. While it won't climb the walls in your home, you can style it as if it would. Make sure to give the large foliage which demands space plenty of room to grow outwards and upwards. Using wire to tie the plant to a wall can help, or, as you've seen me do, place it high in a hanging clay or terracotta planter. To keep it looking its best, remove any dried leaves and wipe the leaves with a damp cloth. Removing the dust that has accumulated on the foliage will bring more light to the tissue of the foliage and give your plant back its natural shine.

SPINDLE PALM

(*HYOPHORBE VERSCHAFFELTII*)

The spindle palm is one of my favorite palms to bring indoors because of how tall they grow and how relatively easy they are to care for. And yes, I use the word "easy" lightly because I'm not one to believe in easy to care for or "hard to kill" plants. All plants come with a bit of difficulty or should I say needs, which must be met in order to see that plant successfully thrive. The spindle palm is no exception. When given the love and care it needs, just watch as this gorgeous palm unfurls fronds that can grow up to 15ft (4.5m) indoors. This will make any space feel more lush and tropical. While it can resemble many other indoor palms in its appearance, the difference is clear in its orangish/brown stem and how robust its foliage is. Here are my tips on how to care for it.

LIGHT

Think about this palm as if you're considering its needs in the wild. For the spindle palm, bright indirect light to morning direct sun would be best. So finding a spot for it in an eastern- or southern-facing window would be ideal. While they tolerate medium light, I'd suggest bringing in a different kind of plant if you don't have the bright light needed to allow it to thrive. The one thing that you want to stay away from is afternoon direct sun, which can burn the foliage, causing the plant to dry out and die over time. Because of how full and lush it can grow, be aware that it may block light from surrounding plants.

WATER

Unlike most palms, the spindle palm needs its soil to be consistently moist. This isn't a palm that will be forgiving when you miss a watering. The best way to keep that soil evenly moist is to make sure you test the moisture level daily. To do so, either stick your finger down in the soil or use the cake test by inserting a chopstick or wooden dowel in the soil. When you pull the dowel out, if it has a bit of dirt or mud on it, it's moist. But if it comes out dry, it's definitely time for a drink. When watering, use

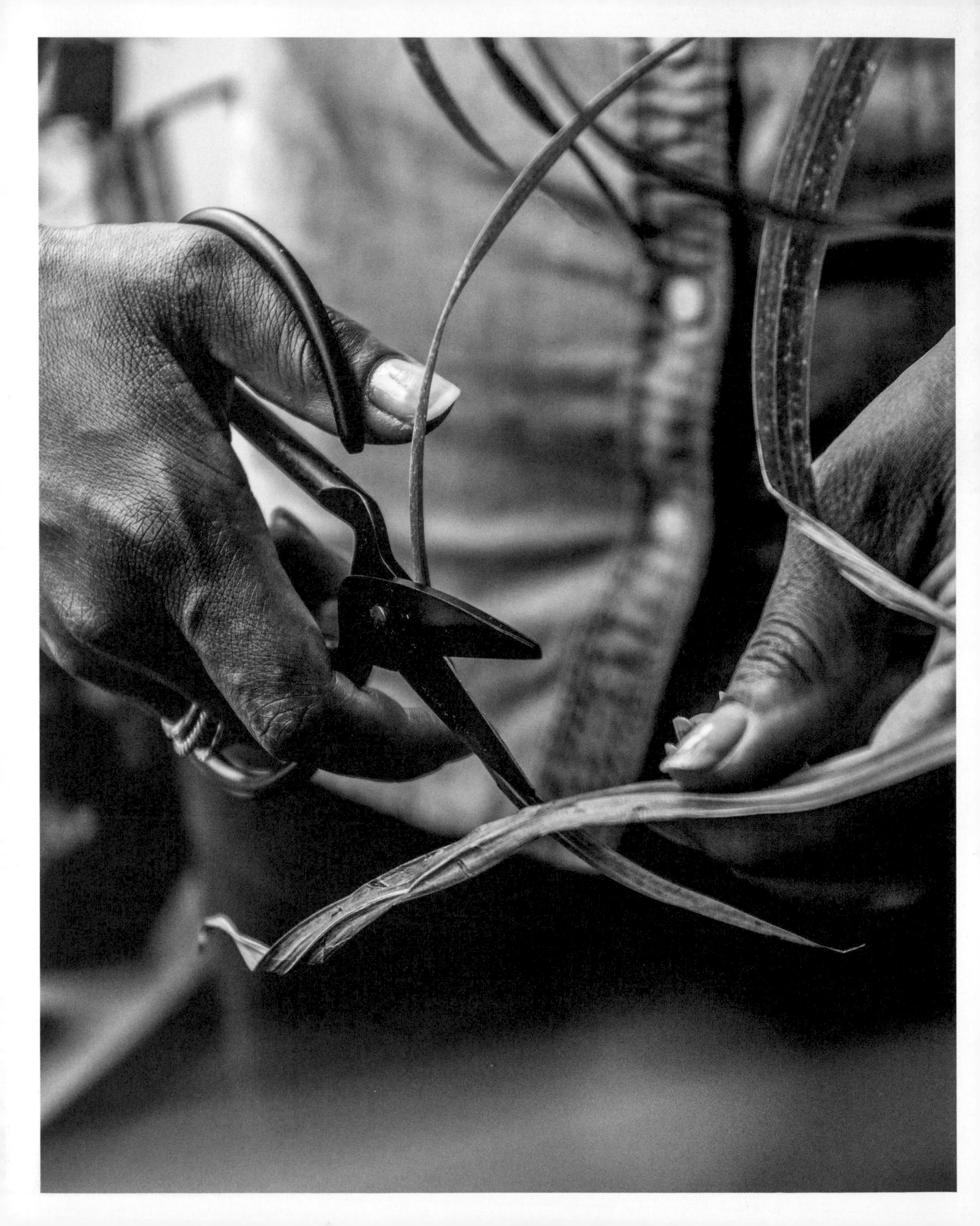

lukewarm water and make sure to water thoroughly until water comes out of the drainage hole of the pot. One of the ways I like to maintain moisture in the soil is to insert a water spike in the soil and place a wine bottle filled with water on top (see *Watering Your Plants While Away* p.112). This will allow moisture to be slowly released throughout the week.

SOIL

The soil of your spindle palm should be well-drained and loose. A mixture of potting soil, sand, and bark will help to make sure it stays aerated (see *Creating Your Potting Own Mix* p.152). To further help aerate the soil, before watering take a wooden dowel and poke holes into the soil. This will help the roots to breathe and water to pass through easily.

REPOTTING

Repot this plant only when the roots start to emerge from the drainage holes of the pot or when you notice the roots becoming bound. This may be every year or every other year. When doing so, make sure to place it in a pot that is 2in (5cm) larger in diameter than the previous pot.

TROUBLESHOOTING

YELLOWING LEAVES This is a sign that your plant is being overwatered. While you're keeping the soil moist, try not to keep it wet. Making sure your pot has drainage will help with this.

BROWN TIPS OR CRACKS ON THE LEAVES This is a sign of underwatering. Check the soil to see if it's dry. If it is, water it right away. Sometimes brown tips are due to moisture not making it all the way to the end of the leaf. To help with the humidity this plant needs, mist it every few days.

BROWN SPOTS IN THE INTERIOR OF THE LEAVES This is a sign of root rot. Take your plant out of its pot and check the soil and roots. The roots should look off-white and sturdy, like an *al dente* noodle. If any look dark brown and mushy, that is a rotting root. Cut them back with your shears and repot in fresh, fast-drying soil.

STYLING TIPS

I've always leaned towards the phrase, "Go big or go home." So why not just go big AT home by placing your palm high up in your space. This not only gives it more presence, but also draws the viewer's eye up, making them more aware of what's above them. For this plant to retain the moisture it needs, I'd suggest styling it in a glazed ceramic or plastic container. When bringing a plant like this indoors, whether it's into your home or office, always consider its growth potential, as you won't want it limiting the space you have to move around in. Because of its sturdy growth, it does well in high-foot-traffic areas.

STAGHORN FERN

(*PLATYCERIUM BIFURCATUM*)

Back in 2011 I took a trip to Terrain, a plant store in Glen Mills, Pennsylvania, and was forever changed. What I saw there left me feeling inspired and curious. One, I then dreamed of having a home that was as lush as the garden café they had there, and two, I wondered, what the hell was that weird plant they had hanging over their tables. At the time I wasn't well versed in plants, so I can't really pretend that not knowing how to identify a plant was a rare occurrence. It was such a strange-looking plant that it made a deep impression on me and I never forgot about it. I would later come to learn that it was the gorgeous staghorn fern. And three years later it would be the second plant purchase I ever made. What I love about this tropical beauty is that in the wild it can stretch its antlers out to about 6ft (1.8m). And the crazy thing is that it does all of this growing on the sides of trees, like an air plant (*Tillandsia*). Yes, like an air plant. Just like air plants, staghorn ferns are epiphytes, which means they are plants that grow on other plants, using them as a host. Wild! To me they are one of the most interesting and enjoyable plants to have as a part of your plant family.

It should be clear why it's coined the staghorn. If not obvious enough, it's due to their leaves or should we say fronds, which take on the appearance of deer antlers. When looking at the staghorn fern, it's important to understand what you're looking at so that you can better care for it. The staghorn fern is one of those plants that can confuse even the most plant savvy of us. The body of this fern is broken up into three parts: the antler fronds, the shield fronds, and the root ball. Taking it layer by layer, first you have the root ball which is where all the roots of that plant gather. They are shallow roots, so they don't require much room to grow indoors, but in the wild they'll be found attached to trees. The shallower the roots, the more susceptible they are to root rot. On top of the root ball you'll notice the fern's shield fronds. These fronds help to protect the roots and hold moisture and nutrients for the plant. As the shield fronds grow, they start off

LIGHT

When it comes to light, treat your staghorn like most ferns, and style it in a spot that gets anything from bright indirect light to indirect light. The more you can expose it to open sky, without having it kissed by direct sun, the better. Dappled light is what it's used to in the wild, but just because it might be in a shadier spot outdoors doesn't equate to it loving a shadier spot in your home. The brighter, the better. But please refrain from placing your fern in direct sun. Direct sun will burn the fronds of your fern, causing them to wilt and dry up quickly, making it more difficult for you to give your plant the moisture it needs. If you have northern or eastern exposure, this would be the perfect spot for a plant like this.

green, like your typical plant leaf, but as they mature, they become brown and hard. When I first saw this happen to my fern I freaked out and thought I had done something wrong. After talking with an assistant at the plant store where I had got it, I learned that it was totally normal. This process helps build a stronger protection and hold to the host plant or board that the fern is mounted on. So please never try to prune back a browning shield frond.

Finally, the stars of the show, the antler fronds, grow from the center of the plant, surrounded by its shield fronds.

I've seen many people display their staghorn ferns mounted on board, but they also do really well in hanging baskets. While their beauty can pull you in and make you want to bring them into your indoor oasis, understanding how to care for them is key. So here are my tips on how to care for the staghorn fern.

WATER

The staghorn fern is a tropical plant, so the first thing you should think about is humidity. Humidity and moisture are key for tropical plants like this and getting that right indoors can be tricky. To keep this fern thriving, you'll have to be extremely attentive, but not on a daily basis. You don't want to helicopter parent the staghorn. They are sensitive to being overwatered. Once a week during the warmer months and once every two weeks during the colder months, you'll want to take your staghorn down from wherever you have it displayed and submerge it in a sink or tub full of lukewarm water. If mounted on a board, flip the board over with the fern side down, and let it float in the water for about 20 minutes. Once it has had a good soak, take it out and let it drain before placing it back in its spot. Submerging it is the best way to make sure all parts of the plant get the moisture they need. At least twice throughout the week, to increase the humidity around your fern, you'll want to mist it with lukewarm water, making sure to get the underside of the antler fronds so that the moisture falls down the plant and into its root. Mist that builds up on the surface of the frond could lead to bacteria building up.

REPOTTING

This isn't really a thing for your staghorn fern because once they are mature enough, they should be mounted on bark or a piece of board. Once their shield fronds start to creep out to the edge of the board, you'll want to attach a larger board to that board. Don't remove your fern from the board that it's already attached to—removing a mounted staghorn from a board can kill the plant.

TROUBLESHOOTING

YELLOWING FRONDS This is a clear sign that the plant is being overwatered. A good rule of "green" thumb when it comes to the staghorn fern is to watch the stiffness of the antler fronds. If the fronds are upright and stiff, it's well-watered. If they start to droop a bit, it's time for a drink.

SMALL LEAVES AFTER DEVELOPING LARGE FRONDS This is a sign that your fern isn't getting enough bright light. Move it into a brighter spot in your home and you'll start to see larger leaves once again.

BROWNING OR WILTING This is a sign of an underwatered or dry plant. This could be due to the fact that it's getting too much direct sun or it's just not being watered as consistently as it should.

STYLING TIP

Mounting staghorn ferns on boards is the best way to display them. Finding a spot for one (or a few) on your gallery wall will bring your art to life. See *Living Art: How to Make a Wall Mounted Plant* on p.12 to learn how to mount your staghorn.

PHILODENDRON 'XANADU'

Oh the wonderful 'Xanadu'! No, I'm not referring to the estate in the movie *Citizen Kane*, I'm talking about one of my favorite philodendrons to bring indoors. The *Philodendron* 'Xanadu' is a tropical beauty that can instantly make a space feel more lush. While it can resemble its larger cousin, the *Philodendron bipinnatifidum*, with its shiny, deep green, finger-like lobes, the 'Xanadu' leaves don't get as large. When a 'Xanadu' is younger, these lobes aren't as pronounced but as the plant matures, the foliage grows larger, pushing the lobes to stretch out further from the leaf, making it look a little more like the leaf of the *P. bipinnatifidum*. Unlike this plant and other philodendrons out there, the Xanadu isn't a climbing plant, meaning that it doesn't use its aerial roots to climb up other plants and surfaces. In the wild, you'll find the 'Xanadu' growing on the jungle floor, with its leaves protruding from its woody, trunk-like arms stretching out for an embrace, creating its bushy shape. This shape gives it a strong presence, making it a true statement plant once mature. While they are perfect growers for the outdoors, if you live in warmer/more humid climates, finding the right spot for them indoors can make anyone feel like they're on a tropical getaway. Here are my tips on how to care for it indoors.

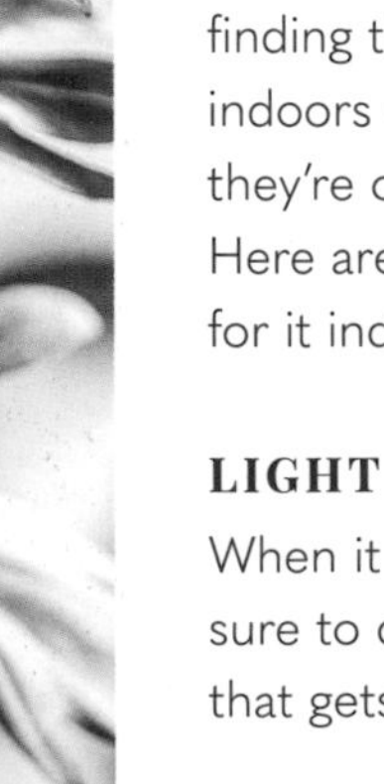

LIGHT

When it comes to light, make sure to dedicate a spot for it that gets anything from bright indirect light to indirect light. While they can tolerate moments of direct sun, too much can burn the foliage, causing the plant to develop brownish-orange spots, and can eventually kill the plant. Find a spot that has a lot of exposure to open sky or a room that filters direct sun into bright indirect light. Because the 'Xanadu' grows in a bush shape, it's important to rotate this plant often for balanced growth.

WATER

Like all philodendrons, you want to make sure they aren't overwatered. In the case of the 'Xanadu', you're better off underwatering than overwatering. The best way to keep your plant healthy and thriving would be to water when the top 2in (5cm) of the soil is completely dry.

To check this I recommend the cake test. Basically stick a skewer or chopstick about 2in (5cm) into the soil and if it comes up muddy, the plant is not thirsty, but if it comes up dry, then it's time to give it a drink. And make sure to only use lukewarm water. Pour the water slowly and evenly across the top of the soil, making sure that all sides of the pot and the roots below can have a drink. You should continue to water your plant until you see water exiting out of the drainage hole and into the base tray. Once that happens, let the runoff water sit in the base tray for about 15–30 minutes, just in case the roots and soil didn't get a chance to hold on to that water as it came rushing down. After 30 minutes, if there is still water in the base tray, take your plant off that base tray, dispose of that runoff water, and then place your plant back in the tray. Given that it's a tropical plant, you'll want to create the climate in your home that it can best thrive in. This means bringing in some humidity. For the 'Xanadu', during the warmer months I recommend misting it at least twice a week and during the colder months, mist every morning with lukewarm water. I've become a huge fan of moisture meters for those out there with more than 50 plants. Sticking your fingers in soil every day/week can become a bit much over time.

SOIL

For your 'Xanadu' and other philodendrons, use a potting mix that is well-aerated and dries appropriately. In this case, having additives in your potting mix that can help pull moisture away from the soil so that the roots aren't sitting in water for long is important. Depending on the planter, I like to create a potting mix that is 80% soil and 20% perlite for my philodendrons. This would be my ratio if I was placing the plant in a clay or terracotta pot, but if it was going in a glazed ceramic pot or plastic container, I would add more perlite or even vermiculite. (See *Creating Your Own Potting Mix* p.152.)

REPOTTING

Only repot your 'Xanadu' when the roots of the plant start making their way out of the drainage hole. The more the roots grow out of the base of the pot, the more susceptible they will be to rotting. Make sure to repot in a planter that is 2in (5cm) larger in diameter than the previous pot. Consider the material the pot is made out of because this will be a factor in the moisture level of the soil. Repot your plants during spring and summer for the best results.

PROPAGATION

The 'Xanadu', unlike its other philodendron friends, can't be propagated using the stem-cut method. For this plant you have to use the division and separation method, which is done by separating sections of the plant at the roots and then moving these to a new pot. Propagate your plants during spring and summer for the best results.

TROUBLESHOOTING

YELLOWING LEAVES This is a sign that your plant is being overwatered. If you've been following the tips on how to water but are still seeing yellowing leaves, you should use a wooden dowel or chopstick and make holes in the soil to aerate it, and if that doesn't help, pull the plant out of the pot and check the soil. If the soil is staying wet for too long this could the reason.

BROWN TIPS OR CRACKS ON THE LEAVES This is a sign of underwatering. Remember to check the soil and make sure that when you do water your plant, that water runs through all of the soil. This could also be a sign of too much fertilizer.

BROWN SPOTS IN THE INTERIOR OF THE LEAVES This is a sign of root rot. Take your plant out of its pot and check the soil and roots. The roots should look off-white and sturdy, like an *al dente* noodle. If any look dark brown and mushy, that is a rotting root. Cut them back with your shears and repot in fresh, fast-drying soil.

STYLING TIP

The 'Xanadu' is a true statement plant. Find a spot for it that gets it off the floor and makes it more of a conversation piece. Try placing your plant in the center of a dining table or on a pedestal. To keep it looking its best, remove any dried leaves and wipe the leaves with a damp cloth. Removing the dust that has accumulated on the foliage will bring more light to the tissue of the foliage and give your plant back its natural shine.

JUNGLE BY NUMBERS MURAL LINE ART (see p.92)

Use this template or print the image out from this website: thingsbyhc.com/junglebynumbers

INDEX

CREDITS

JAMIE CAMPBELL
@jamiecampbellbynum

JASMEN DAVIS
@jasmendavis

LETTA MOORE (p.21)
@ksmcandleco

SARA TOMKO (p.30)
@hideandpeak
hideandpeak.com

MATT NORRIS (p.56)
@anatomatty

DRURY BYNUM (p.93)
@drurybynum

CERAMICISTS FEATURED:

DANA BECHERT (p.114)
@danabechert

KRISTINA BING (p.141)
@kbing

CLAIRE DI SALVO (p.28)
@milkweedceramics

HOMEBODY GENERAL (p.46)
@homebodygeneral

C O R B É (p.185)
@corbecompany

WHITNEY SIMPKINS (p.162)
@personalbestceramics

THANKS AND LOVE

I feel so incredibly thankful to have had the opportunity to create this book. With everything happening this year, it's almost a miracle that it even came about. That miracle I speak of was just the team I had around me to help push it all through. Without them, there wouldn't be a *Wild Creations* and I want to take this time to thank them.

As always, I'd like to start by thanking my amazing wife, Fiona. She's the source of my drive and my happiness. Every idea I've had over the past five years I've bounced off of her and as it made its way back to me, it returned more profound and polished. Fiona, you're the tailor of the smile I wear daily. I've been made stronger, smarter, and whole because of you and I continue to grow and thrive in response to your care. Without you, none of this would be possible. I love you deeply.

Thank you to my loving mother, who has always been the cheerleader on my side. I hope I continue to become your dream realized. To my close friends who have always been there to support me and have helped shape the individual I am today, thank you.

Thank you to all of the wonderful individuals that worked to make these projects come together. I'd like to start by thanking Jamie Campbell, who worked as a stylist on each project, and on some, a hand model. Jamie, thank you for remaining patient with me throughout the production of this book. Thank you for putting in so much hard work and challenging me to make each project better. This book is beautiful because of your help. Thank you to Jasmen Davis, who helped as a hand model and assistant for many of the projects. Jasmen, you really made the studio more lively and fun. We couldn't have done so much without your hands helping to carry the weight. To Letta Moore, thank you for showing us all how to create beautiful candles to set the vibes off right in our spaces. You've created something truly special at KSM and I look forward to watching you thrive. To Sara Tomko, thank you for helping me create the hanging plant stand I've always wanted. Your guidance and style made something special. To Matt Norris, thank you for always making yourself available whenever I have an idea that involves woodworking. You're a real craftsman and a great collaborator. And finally to Drury Bynum, thank you not only for helping all of us create jungle scenes in our homes but also creating the beautiful cover art for the book. You're one of the most prolific individuals I know and I'm thankful to have you as a friend.

Thank you to the team at CICO books for working with me again and again. To Cindy, over the years we've cultivated a wonderful working relationship and it's your faith in me that keeps this train in motion. Thank you. To the team there that help make all my wild ideas make sense on paper, Megan and Martha, thank you. As always, the entire lot of you put so much time and energy into making this book and I'll be forever grateful.

Lastly, to you, the green-loving community, once more, thank you. It been such a wild ride over the past few years and because of your continued support, this opportunity was created. I hope that within the pages of *Wild Creations* you find a bit of inspiration to create your very own wild interior. I hope that you'll have a bit of fun creating the projects, find use for the hacks, and find a piece of wisdom in my rants. I wish you all nothing but the very best and greener plants. With that said, keep dancing in the dappled light and always stay wild!

This book is dedicated to all of the creatives out there. Keep pushing.